The Δa photometric system

The Δa photometric system

Ernst Paunzen

Department of Theoretical Physics and Astrophysics
Masaryk University
Kotlářská 2
CZ - 611 37 Brno
Czech Republic

Bibliographical information held by the German National Library

The German National Library has listed this book in the Deutsche Nationalbibliografie (German national bibliography); detailed bibliographic information is available online at http://dnb.d-nb.de.

1st edition - Göttingen: Cuvillier, 2020

Nonnenstieg 8, 37075 Göttingen, Germany

Telephone: +49 (0)551-54724-0

Telefax: +49 (0)551-54724-21

www.cuvillier.de

1st edition, 2020

This publication is printed on acid-free paper.

ISBN 978-3-7369-7283-4

eISBN 978-3-7369-7283-5

Contents

1 Summary

This book describes the history, characteristics, observations, results and future prospects of the Δa photometric system. Up to now, it successfully produced countless scientific output since its development and first appearance on the astronomical scene initiated by Hans-Michael Maitzen in 1976.

It was intended to study the typical and unique flux depression at 5200 Å found in mainly magnetic chemically peculiar stars of the upper main sequence. Therefore the most important measurement is through the g_2 filter centred at this wavelength region. In addition, one needs the information about the continuum flux of the same object. Originally, the g_1 (centred at 5000Å) and Strömgren y (5500Å) was used to get the continuum flux. In principle, any other filters such as Johnson $BVRI$ can be used. However, the closer it is measured to the 5200Å region and the narrower the used filters are, the results become more accurate. The a index is then defined as the flux in g_2 compared to the continuum one. Because there is a general increase of opacity around 5200Å with decreasing temperature, one has to normalize a with the index a_0 of a non-peculiar star of the same temperature, to compare different peculiar (or deviating) stars with each other (Δa index).

With this photometric system it is possible to measure any peculiarities, such as abnormal absorption and emission lines, in the region of 5200Å. Soon, it was used to investigate metal-weak, Be/shell and supergiants in the Galactic field and open clusters.

At the end of the 1980ies, the highly developed photomultiplier technology, slowly but surely, was replaced by the CCD technology. It took a few years until CCDs reached the same accuracy as photomultipliers for the same exposure time and target brightness.

The first CCD Δa observations which were performed 1995 with the same telescope, the 61 cm Bochum telescope at La Silla, by the same person, Hans-Michael Maitzen, who initiated in 1971 the observations of CS Vir in g_1 and g_2. The results of the first CCD observations, data for 25 Galactic field stars with known spectroscopic peculiarity types, were published in 1997. After that, CCD observations of open clusters, the Large Magellanic Cloud and Globular clusters were successfully secured.

This book goes one step further showing that the Δa photometric system and the study of its corresponding spectral region from 4900 to 5700Å is able to contribute significant new insights at many fields of astrophysics. The list of objects which are potential targets is long: cool-type Population

I and Population II objects, supergiants, emission type objects, all type of galaxies, and so on.

These research fields can be investigated by 1) new observations; 2) archival data not used so far, for example cool-type objects in star cluster fields and 3) data from other surveys which employ similar filters. For sure, all three approaches will be extensively utilized in the future.

2 The early history of the Δa photometric system

The roots of the Δa photometric system date back to the year 1969. The invention and development of it is closely tied to ao. Univ.-Prof. i.R. tit. Univ.-Prof. Dr. Hans-Michael Maitzen (HMM). The interested reader is referred to his article "Two decades of Δa photometry" published by Maitzen (1998) which includes a comprehensive overview of the history until then.

Here, a very short overview is given because it sheds more light on the original intentions which are still state-of-the-art today.

In 1969, HMM got a position at the newly founded astronomical institute of the Ruhr-University of Bochum, Germany. Soon, the at that time, head of the institute, the late Prof. Theodor Schmidt-Kaler invited him to perform photometric observations at the Bochum 61 cm telescope at ESO (La Silla). He started the observations in May 1969, only less than two months after the official opening of the observatory on 23. March 1969. He monitored magnetic stars (CP2) from the catalogue by Babcock (1958) as suggested to him by the late Prof. Karl Rakos. This was to confirm the variability of these stars by means of Johnson UBV photometry. The "key star" for the further historical development was CS Vir (HD 125248), a classical CP2 object. HMM found not only, as expected, a decrease of the variability amplitude from U to B, but also the V light curve exhibited either half the period of U and B, or as we know today, a double wave variation due to the variable continuum backwarming produced by strong line absorption at shorted wavelengths.

These results were published by Maitzen & Rakos (1970). As next step, photometric observations in the Strömgren system and spectroscopy of CS Vir was done. Because no filter covered the wavelength interval where the transition from a single to a double wave variation occurred, he decided to order two new interference filters (g_1, $\lambda_c = 5020$ Å and g_2, $\lambda_c = 5240$ Å, both with FWHM $= 130$ Å). These filters are already very close to those used today (Sects. 9 and 14.1.1). From their colour appearance to the human eye, Prof. Schmidt-Kaler suggested to name them *giftgrün* (poisonous green, g_1) and *lindgrün* (yellowish green, g_2) which was rejected for practical reasons. The first observations with these filter were conducted in 1971 in order to trace the light variability of CS Vir but not in the way, the Δa photometric system is used nowadays. The most important result for the birth of the system was the detection of the characteristic flux depression at 5200Å by Kodaira (1969) which can be measured via

g_2. This prompted HMM to substantiate the feature by a more extended photometric campaign of different CP stars (not only magnetic ones) and apparent normal type ones. Using the already available Strömgren y filter, he defined the normality line, and the final Δa parameter. The foundation paper then appeared more than 40 years ago (Maitzen, 1976a).

3 The diagnostic tools of the Δa photometric system

Here, an overview of the characteristics and diagnostic tools of the Δa photometric system is given. One has to refer to many consecutive Sections where these tools are applied and discussed in more detail. Some of the information given here might be repeated there because of the necessity to clarify some points within the context. However, it was tried to minimize such duplicities.

Basically, the Δa photometric system consists of one filter which measures the 5200Å region (g_2) and an additional information about the continuum flux of the same object. This can be either achieved by measuring the flux at the adjacent spectral regions, for example with the filters g_1 (5000Å) and $g_3 = y$ (5500Å), but also by any other effective temperature (T_{eff}) sensitive colour, $B-V$, $b-y$, or $B2-V1$ to give some examples. The latter were indeed successfully applied, for example, Maitzen et al. (1986) used $b-y$, whereas Maitzen (1993) also employed $B-V$ colours.

With these measurements, the index a can be calculated as

$$a = g_2 - \frac{g_1 + y}{2} \quad (1)$$

$$a = g_2 - \frac{b + y}{2} \quad (2)$$

$$a = g_2 - \frac{B + V}{2} \quad (3)$$

In principle, the positioning of the continuum filters minimizes the influence of T_{eff} on the peculiarity index. The closer it is measured to the 5200Å region and the narrower the used filters are, the results become more accurate. One has to avoid, for example, the hydrogen or helium lines for the continuum measurement because they are very much sensitive to the projected rotational velocity and the surface gravity.

In Sect. 10, Table 9 lists several different filters which were used in the past for different purposes to obtain observations. There is a general increase of opacity around 5200Å with decreasing temperature (Sect. 14.1).

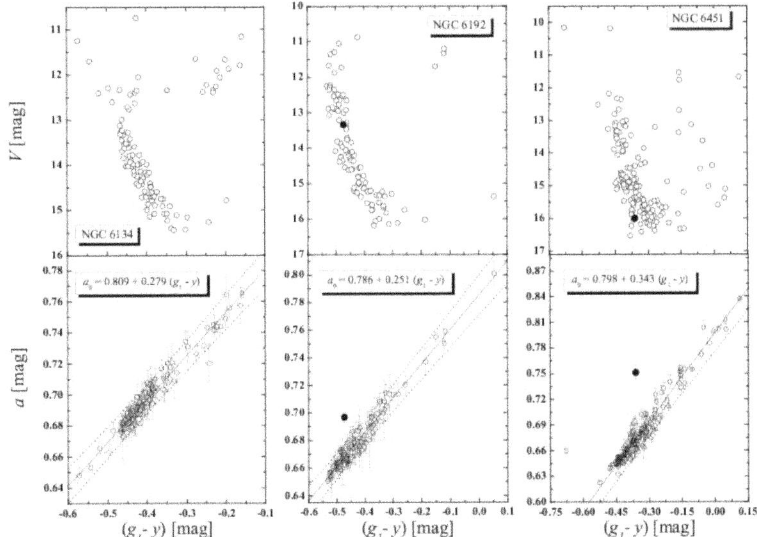

Figure 1: The two basis diagnostic tools of the Δa photometric system for three open clusters: the normality line a_0 versus $(g_1 - y)$ and colour-magnitude M_V versus $(g_1 - y)$ diagrams together with isochrones.

Therefore, one has to normalize a with the index a_0 of a non-peculiar star of the same temperature, to compare different peculiar (or deviating) stars with each other. The photometric peculiarity Δa index is therefore defined as

$$\Delta a = a - a_0[(g_1 - y); (b - y); (B - V))]. \tag{4}$$

The first diagnostic tool is the *normality line* which comprises of the location of the a_0-values in respect to a T_{eff} sensitive colour. Assuming that all stars exhibit the same interstellar reddening, peculiar objects deviate from the normality line more than 3σ (Fig. 1).

From observations, it is known that the normality line is shifted by $E(b-y)$ to the red and by a small amount $E(a)$ to higher a-values (Maitzen, 1993). It has to be emphasized that correlations between the amount of reddening for different photometric systems exist (Sect. 12.1.2, Yuan et al., 2013), for example $A_V = 3.1E(B - V) = 4.3E(b - y)$. These correlations can be used to transform the reddening values between the different photometric systems. The ratio of the shifts

$$f = E(a)/E(b - y) \tag{5}$$

can be determined from the deviation of a reddened cluster normality line from the unreddened relationship. On the other hand, assuming a mean f-value (≈ 0.05) and iterating the formula

$$a(\mathrm{corr}) = a(\mathrm{obs}) - fE(b-y) \tag{6}$$

one can determine reddening values by the Δa-photometry of clusters. The effect has only to be taken into account for $E(b-y) > 0.3$ mag and a non uniform reddening distribution.

More of a problem is the estimation of differential reddening within a star cluster (Bonatto et al., 2012), for example. In principle, several methods can be applied (Sects. 12.1.2 and 12.2.2) to overcome this problem:

- to deredden each individual object using the Strömgren $uvby\beta$ photometric system and its calibrations
- to use the Q-method for the Johnson UBV photometric system

An a versus $(g_1 - y)_0$ or $(b-y)_0$ or $(B-V)_0$ diagram should then be able to further strengthen the membership of objects to the investigated star cluster since fore- and background deviate significantly from the normality line. These deviations are in generally ten times higher than those observed for the most prominent CP stars.

The second diagnostic tool is a y versus $(g_1 - y)$ diagram. Instead of y, any other magnitude, for example, V, R, I, and so on, can be taken. This is just a classical colour-magnitude diagram (CMD) for the Δa photometric system. With the knowledge of the reddening and the distance, the absolute magnitude M_V, and using the bolometric correction (B.C.) as well as the absolute magnitude of the Sun, the luminosity $\log L/L_\odot$ can be calculated (Gómez et al., 1998).

In Sect. 13, a grid of isochrones with different initial chemical compositions for the Δa system is presented. There, it is shown that the accuracy of fitting isochrones to Δa data without the knowledge of the cluster parameters is between 5 and 15 %. This efficient tool has been already widely used for star clusters and the LMC (Sects. 17 and 19).

Figure 1 shows M_V versus $(g_1 - y)$ diagrams (upper panels) together with isochrones for three open clusters. The capability to sort out non-members and to estimate the cluster parameters are evident.

4 Other indices measuring the 5200Å region

One has to differentiate between indices which measure the flux in the 5200Å region directly and those which measure the metallicity (line blan-

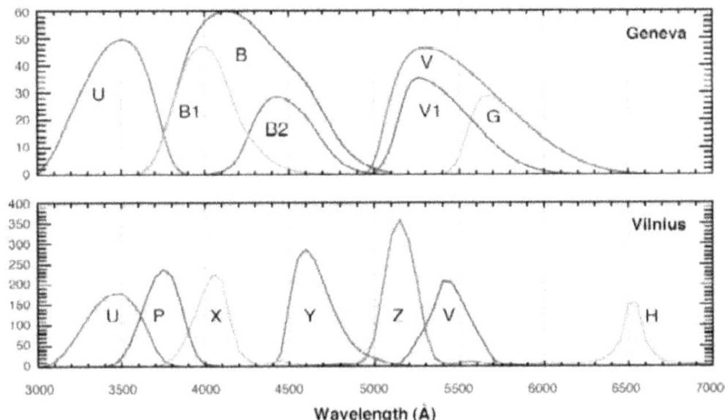

Figure 2: The filter transmission curves of the Geneva 7-colour (upper panel) and Vilnius (lower panel) system taken from Bessell (2005).

keting). The latter are capable to detect most extreme CP stars because of their prominent overabundances. Those are, for example, the m_1, Δm_1, and $[m_1]$ in the Strömgren system as well as m_2 and δm_2 in the Geneva 7-colour system (Golay, 1974).

Another interesting approach was published by Masana et al. (1998). They defined a reddening-depended peculiarity index Δp as a linear combination of all Strömgren $uvby\beta$ colours. They found a detection rate of up to 50% for early (hot) CP2 stars. However, this approach was not systematically followed after the original paper has been published. It would be very interesting to use the new photometric catalogue by Paunzen (2015) to update and upgrade this method.

Besides the Δa photometric system (Sect. 3), a few other indices are available which measure the flux in the 5200Å region. These are described in more detail in the following.

The $\Delta a'$ index by Adelman (1979): it basically works in the same way as the "classical Δa index", but was especially designed for his spectrophotometric data. It is defined as

$$a' = m_{5264} - [m_{4785} + 0.453 \cdot (m_{5840} - m_{4785})]. \tag{7}$$

The measurements at 4785 and 5840Å, which were both made at locations significantly bluer than g_1 and redder than y, represent a wavelength base for the continuum that is larger by a factor of two than in Δa photometry. The depression itself is represented by the 5264Å value. Like the a index, the a' index is slightly dependent on the colour of the star and therefore has to be normalized by the index of non-peculiar stars with the same colour.

The $\Delta(V1-G)$ and Z indices of the Geneva 7-colour system: The Geneva 7-colour photometric system (Fig. 2) is the most homogeneous one because unique filter sets together with the same type of photomultipliers were used throughout its history (Golay, 1994). However, unfortunately, no new observations are available for it because the original instrumentation is not existing any more. The $\Delta(V1-G)$ index is the first measurement for peculiarity derived from the Geneva 7-colour photometric system. The $V1$ and G filters are centred at 5408 and 5814Å (bandwidths of about 200Å), respectively. Hauck (1974) was the first to propose this index as peculiarity parameter. It is defined as

$$\Delta(V1-G) = (V1-G) - 0.289 \cdot (B2-G) + 0.302. \qquad (8)$$

On average, normal stars have $\Delta(V1-G)$ values of -5 mmag (Sect. 11.11). The zero point of the $\Delta(V1-G)$ index represents the upper limit of the sequence of normal type objects and not its mean value. This was done by using the upper envelope for normal type, luminosity class V to III objects, based on a linear fit for the correlation of $(V1-G)$ with $(B2-G)$ as given by Hauck (1974) which introduces this negative shift. Consequently, a very strict significance limit of $+10$ mmag for $\Delta(V1-G)$ was set by Hauck & North (1982) to avoid contamination of CP objects.

Besides the $\Delta(V1-G)$ index, Z within the Geneva 7-colour photometric system is most suitable for detecting CP stars. Originally, Cramer & Maeder (1979) defined a three-dimensional grid $[XYZ]$ for hot stars on the basis of the seven available filters (Fig. 2). The X parameter is oriented along the main sequence (MS) of O- and B-type, whereas the Y is in the direction of high luminosity stars. The Z parameters is normal to the $[XY]$ plane. It is defined (Cramer, 1999) as

$$Z = -0.4572 + 0.0255 \cdot U - 0.1740 \cdot B1 + 0.4696 \cdot B2 - 1.1205 \cdot V1 + 0.7994 \cdot G \qquad (9)$$

The Z index is virtually independent of temperature and gravity effects for stars hotter than A0 or $(b-y)_0 < 0$ mag. Furthermore, the only stars showing a significant deviation in the Z direction, are the CP stars. Later on, Cramer & Maeder (1980) showed that Z is correlated with the measured magnetic field of CP stars. This is line with the corresponding synthesized photometry (Sect. 14.9.2). Cramer (1999) lists a limit of ± 10 mmag for apparent peculiarity (Sect. 11.11).

As the Δa index (Sect. 5.1), $\Delta(V1-G)$ and Z also vary over the rotational period of CP2 stars (Muciek et al., 1984). However, the variations (± 3 mmag) are less than those for the Δa index.

The different diagrams of the Vilnius system: The Vilnius system (Fig. 2) was developed independently from the Geneva 7-colour system but for similar reasons, namely, to derive temperatures, luminosities, and peculiarities in reddening and composition from photometry alone. The colours are normalized by the condition $U-P = P-X = X-Y = Y-Z = Z-V = V-S$ for unreddened O-type stars. Therefore, all colours for normal stars are positive. Reddening free indices are constructed as for the Geneva 7-colour and Strömgren systems. Interesting enough, the Z filter was placed on the Mg I triplet as well as the MgH molecular band. It is also sensitive to the luminosity classes of G- to M-type stars (Straižys et al., 1986). Straižys & Žitkevičius (1977) proposed four diagrams to separate CP from normal type stars. They defined the following index

$$Q_{YZS} = Y - Z - 0.538 \cdot (Z - S) \tag{10}$$

which is, in principle, independent of interstellar reddening. If a star exhibits $Q_{YZS} < -0.03$ mag it can be considered as CP object. They found that 56% of CP stars (denoted as Ap stars in their work) have such values with the classical hot Si stars show the largest and the HgMn (CP3) the smallest deviations. It turned out that strongly reddened and cool type CP stars can not be unambiguously detected. Also, this index fails to detected CP1 stars which is similar to the situation within the Δa photometric system (Sect. 11.3). However, CP1 stars can be detected in the Vilnius system on the basis of CMDs including the U filter. Later on, North et al. (1982) discussed the capabilities of a joint Vilnius and Geneva 7-colour (VILGEN) system to detect CP stars. They improved the detection capability of $\Delta(V1-G)$ by using the Vilnius Z filter (not to be confused with the Geneva Z index) instead of $V1$. In addition, they defined four peculiarity indices (PECx). Using these indices, 54% of magnetic Ap (CP2 and CP4) stars were detected. Since then, no new corresponding investigation has been published employing this extended system.

5 The variability of the Δa index

If one looks very closely on the brightness of any arbitrary chosen star, one will find variability on different time scales. The detection limit of the amplitude only depends on the instrumentation and accuracy of the measurements. The Sun, as the closest star, shows variability with periods from minutes (not visible by eye) to several years (sunspots). Variable stars are divided into two categories: intrinsic variables, in which internal physical changes, such as pulsations or eruptions, are the driving mechanism, and extrinsic variables, in which the light output fluctuates due to planet transits, eclipses or stellar rotation (Percy, 2007). The further classification is rather complex; originally it was based on a star's light-curve characteristics, amplitude, and periodicity (or the lack of it). Many astrophysical theories, for example pulsation, diffusion, and evolutionary models, can be tested with variable stars (Handler, 2013).

It is also well known that the amplitude of almost all types of variability is depending on the observed wavelength region. For example, CP stars exhibit larger amplitudes in the UV than in the IR region (Krtička et al., 2012). Therefore, it was several times tested if and how the variability of stars influences the 5200Å region and the Δa index. The knowledge of this behaviour is essential to estimate the percentage of missed positive (or negative) detections, for example.

5.1 CP stars

Here, the published results of the analysis of the variability behaviour for 17 well established CP stars is presented. Let us recall that the main characteristics of this group are (Sect. 6): peculiar and often variable line strengths, quadrature of line variability with radial velocity changes, photometric variability with the same periodicity and coincidence of extrema. Stibbs (1950) introduced the Oblique Rotator concept of slowly rotating stars with non-coincidence of the magnetic and rotational axes. This model reproduces the photometric variability by the appearing and receding patches on the stellar surface similar to Sun spots.

Table 1 lists the spectral types (Skiff, 2016), dereddened $(b-y)$ colours (Sect. 10), averaged quadratic effective magnetic field $\langle B_e \rangle$ (Bychkov et al., 2009), the rotational period P, the observed Δa range, and the mean absolute Δa value of the sample. It was chosen to list here the $\langle B_e \rangle$ values by Bychkov et al. (2009) because they compiled a homogeneous sample of measurements. For the definition of $\langle B_e \rangle$ and its correlation with the magnetic field modulus, for example, the reader is referred to Bychkov et al. (2003). Briefly, the effective magnetic field B_e of a star is a complex average over the stellar disc of the projection of the local vector of magnetic intensity on the line of sight, and can be measured as the splitting of circularly polarized components of spectral lines.

Table 1: The spectral types (Skiff, 2016), dereddened $(b-y)$ colours (Sect. 10), averaged quadratic effective magnetic field $\langle B_e \rangle$ (Bychkov et al., 2009), the rotational period P (references see text), the observed Δa range (references see text), and the mean absolute Δa value (Sect. 10). For some objects, the mean absolute Δa value could be higher than the given range because there is an offset between different applied filter systems (see Section 7 in Hensberge et al., 1981).

HD	HIP	Spec	$(b-y)_0$ [mag]	$\langle B_e \rangle$ [G]	P [d]	Δa_{range} [mmag]	Δa [mmag]
3980	3277	Ap SrEuCr	+0.079	780(25)	3.9516	+39 ... +49	+38
5601	4488	Ap Si	−0.056	1192(100)	1.11	+52 ... +66	+49
19712	14736	A2p CrEu	−0.060	2268(225)	2.1945	+44 ... +62	+43
25267	18673	B9p Si	−0.070	241(91)	1.21	+30 ... +50	+36
30466	22402	B9p SiCr	+0.025	1716(354)	2.7795	+42 ... +57	+50
30849	22340	Ap SrCrEu	+0.166		15.865	+12 ... +37	+30
38823	27423	A5p Sr	+0.151	1509(111)	8.635	+06 ... +16	+19
50169	32965	Ap SrCrEu	−0.035	1218(70)	1.729	+65 ... +67	+78
52847		Ap CrEuSr	+0.102			+53 ... +58	+62
53116	34049	Ap SrEu	−0.046		11.978	+38 ... +63	+50
55540		Ap CrEu	−0.068			+66 ... +71	+70
56022	34899	Ap SiSr	−0.022	202(117)	0.9184	+18 ... +26	+16
72968	42146	A0p SrCr	−0.039	445(181)	11.305	+12 ... +21	+50
81009	45999	A2p SrCrEuSi	+0.086	1431(204)	33.984	+29 ... +37	+35
111133	62376	A0p SrEuCr	−0.057	807(143)	16.31	+43 ... +73	+56
116458	65522	Ap CrSrEu	−0.036	1926(273)	126.18	+14 ... +32	+54
126515	70553	A2Vp SrSiCr	−0.038	1859(360)	129.95	+33 ... +74	+52

Note that the $g_1, g_2, g_3/y$ filters, for example, used by Hensberge et al. (1981) are slightly different than those used by Maitzen (1976a). Therefore, there is an offset between these systems (see Section 7 in Hensberge et al., 1981). In Sect. 10, mean absolute Δa values from different sources are presented. These values can be, for some objects, higher than those derived by the references mentioned in the following. However, the aim of this analysis is, if the individual CP stars would have been detected independent of the knowledge about the observed phase of variability. This is the most general case possible. Or in other words, is the expected minimum Δa value over the phase still above the detection limit (Sect. 10). In the following, the results of the individual stars, listed in Table 1, are discussed in more detail.

HD 3980: Nesvacil et al. (2012) performed a spectrum analysis in order to determine atmospheric parameters for Doppler imaging. They detected also horizontal inhomogeneities (stratification) in the stellar atmosphere. However, no obvious correlation between theoretical predictions of diffusion in CP stars and the abundance patterns could be found. The rotational period of 3.9516 d was deduced by Maitzen et al. (1980). They also investigated the behaviour of the Δa index over the rotational period (+39 ... +49 mmag).

HD 5601: Besides the classification as CP star, not many investigations were devoted to it. Joshi et al. (2009) unsuccessfully search for rapid oscillations (periods below 25 minutes) whereas Hensberge et al. (1981) estimated a rotational period of close to one day. The Δa index varied for ±7 mmag on a very high absolute level.

HD 19712: The rotational periods published by Hensberge et al. (1981, 2.1945(2) d) and Dubath et al. (2011, 2.0422 d) are slightly different. The latter is based on the Hipparcos photometry (Høg et al., 2000). However, Hensberge et al. (1981) noticed that one of their apparent constant comparison star, HD 20319, seems to be a low amplitude variable. Speckle interferometry (Horch et al., 2006) revealed that HD 20319 is a close binary system with a separation of about 1". The variability of this object could be due to binarity. For HD 19712, the variations of Δa range from +44 to +62 mmag.

HD 25267: Glagolevskij & Nazarenko (2015) presented a detailed analysis of the magnetic field geometry for this star with its period slightly longer than one day. The rather large variability of the Δa index (Hensberge et al., 1981) can be possibly explained by the strong dipole shift across the axis when the magnetic poles are located close to each other (Glagolevskij & Nazarenko, 2015).

HD 30466: It is a double wave photometric variable CP2 star (P = 2.7795 d) with an averaged quadratic effective magnetic field of about 1.5 kG (Bychkov et al., 2009). The measured Δa values over the rotational period are between +42 and +57 mmag (Maitzen, 1976b) which are well above detection limit.

HD 30849: This star was often used as a "standard object" which does not show any rapid oscillation (Balmforth et al., 2001). It is one of the coolest CP star investigated for variations in the Δa index. The rotational period of about 15.865 d (Hensberge et al., 1982) is well established. The lower measured Δa of +12 mmag (Hensberge et al., 1981, 1982) is close to the defined detection limit.

HD 38823: Hensberge et al. (1981) found a rotational period of 8.635 d for this cool CP star with a rather strong averaged quadratic effective magnetic field (1.5 kG, Bychkov et al., 2009). The maximum observed Δa index of +19 mmag (Sect. 10) is not very high, probably because of its rather low T_{eff} and the decreasing sensitivity of Δa in this regime (Sect. 14.9.2). In certain phases of the rotational cycle, HD 38823 would have been not detected photometrically.

HD 50169: There is variability of a very long time scale (at least 1200 d) and one which should be connected to the rotational period (1.729 d) present (Heck et al., 1987). However, none of these periods is firmly settled. Adelman et al. (1998), on the other hand, found no variations in their own Strömgren *uvby* as well as Hipparcos photometry. Therefore, the characteristics of the variability still remains unsolved. But due the well established CP classification (Adelman, 1981) and the measured magnetic field, one should expect stellar spots and rotational induced variability. This conclusion is support be the very large Δa value of +78 mmag (Sect. 10).

HD 52847: Up to now, no evidence for variability of this star was found (Maitzen, 1976a; Hensberge et al., 1981). The measured range of Δa values was interpreted as intrinsic scatter. Maitzen (1976a) deduced $\Delta a = +63(4)$ mmag from 16 measurements whereas Hensberge et al. (1981) presented only three measurements. Freyhammer et al. (2008) confirmed the CP nature of HD 52847 and also found a strong mean magnetic field modulus of 4.4 kG. Based on photometric data from the All Sky Automatic Survey (ASAS, Pojmanski, 2002) they established an upper limit for the amplitude of variability of 4.3 mmag.

HD 53116: Hensberge et al. (1981, 1982) found a rotational period of 11.978 d with evidence for an antiphase relation of variations for Strömgren v with *uby* (Vogt & Faúndez, 1979). The presence of a strong magnetic field was only concluded on the basis of photometric data (North & Kroll, 1989). The significant variations not only in Δa, but also within Strömgren *uvby* strongly supports that a strong magnetic field is present causing large surface spots.

HD 55540: Freyhammer et al. (2008) detected a strong mean magnetic field modulus of 12.7 kG for this object. There are no other investigations of this star available in the literature which shows that many members of this group are still hardly investigated. Maitzen (1976a) list $\Delta a = +70(5)$ mmag and found no signs of variability. Hensberge et al. (1981) reported significant differences (40 mmag) in Strömgren *ub* on a long time base. No other investigations in this respect are available in the literature.

HD 56022: This is a CP2 star with a rather weak averaged quadratic effective magnetic field and a well established period of 0.9184 d. Hempelmann & Schöneich (1988) presented a detailed analysis of its light curve in the UV region including the spot characteristics using data from the TD-1 (Humphries et al., 1976) and ANS (Wesselius et al., 1982) satellite experiments. HD 56022 is one of the rare CP2 stars for which the "null wavelength" λ_0, defined where for $\lambda < \lambda_0$ a light curve is in counterphase to a curve $\lambda > \lambda_0$, is between 2500 and 3300 Å (NUV region). Hensberge et al. (1981) found no significant variations in Strömgren *vby* but for *u*. This is in line with the results by Hempelmann & Schöneich (1988) who found the largest amplitudes in the Far-UV with a decrease to the Near-UV. The absolute Δa value (+16 mmag) and its variations are very small and can be explained by the above mentioned characteristics.

HD 72968: Maitzen et al. (1978) presented extensively photometric data for this double wave variable CP2 star with a period of 11.305 d. The listed observations (see Table 1 therein) were used to calculate Δa over the period, resulting in a range between +12 and +21 mmag. This low values can be explained by the rather weak averaged quadratic effective magnetic field (500 G, Bychkov et al., 2009) of this star. According to the detection limit deduced in Sect. 11, it would be recognised as CP star candidate only in certain phases.

HD 81009: This is a very close visual binary system with a separation of the components ($\Delta V = 1.2$ mag) between 0.1 and 0.2" (Tokovinin et al., 2015). The properties of this very interesting system was investigated in more detail by Wade et al. (2000). They found that HD 81009 is a highly eccentric ($\epsilon = 0.718$), long-period ($P_{\rm orb} = 29.3$ yr) binary system composed of two MS A-type stars. The hotter primary component was identified as the slowly rotating ($P_{\rm rot} = 33.984$ d) magnetic CP2 star. The longitudinal magnetic field strength varies between 6.8 and 9.8 kG. The photometric variations presented by Hensberge et al. (1981) have the highest amplitude in Strömgren *v*. No significant light variations were detected in Strömgren *y* and Δa.

HD 111133: Again, a CP2 star with a moderate strong averaged quadratic effective magnetic field of about 1 kG (Bychkov et al., 2009) and a rotational period of 16.31 d (Buchholz & Maitzen, 1979). The latter investigated this object in the region between 4800 and 5600Å using a spectrum scanner. They observed Δa ten times over the complete phase. The values ranged between +43 and +73 mmag which makes this star to one with the largest measured Δa values.

HD 116458: This is a spectroscopic binary system containing at least one CP star with an orbital period of 126.18 d (Ren & Fu, 2013). There was also a 2 kG strong averaged quadratic effective magnetic field detected (Bychkov et al., 2009). It is a very unusual and peculiar object showing strong variations of Co, Hg, Mn, and Si (Dworetsky et al., 1980). The original mean Δa measurement was +54 mmag (Maitzen, 1976a). Measurements at two different phases by Maitzen & Wood (1977) resulted in a relative Δa value variation between 14 and 32 mmag, respectively.

HD 126515: Glagolevskij (2005) modelled the magnetic field variations

Table 2: The observed magnitude ranges, defined as the difference between the lowest and highest value over the rotational period, for Δa, Strömgren *uvby*, and Johnson *V*, if available. For HD 111133 and HD 126515, no additional data to the Δa ones are available.

HD	Δa [mmag]	u [mmag]	v [mmag]	b [mmag]	y [mmag]	V [mmag]
3980	10	39	85	56	21	
5601	14	54	43			22
19712	18	44	34	30	19	41
25267	20	76	60	29	24	51
30466	15		61	55	31	28
30849	25	51	92			20
38823	10	70	130			24
50169	2	5	6	5	7	4
52847	5	11	11	24	6	8
53116	25	107	69	58	103	56
55540	5	45	9	4	10	17
56022	8				19	
72968	9	34	25	33	35	
81009	8	17	56	18	7	
116458	18	86	67	63	54	

($P = 129.95$ d) assuming a strongly decentred magnetic dipole in order to achieve the best agreement between the calculated and observed phase relations. He found that the field structure is determined by a barlike dipole with a considerable distance between the monopoles. Hensberge et al. (1986) presented a detailed investigation of the Δa variability in correlation with the different magnetic field components. Their most important conclusion was that magnetic intensification effects do not control the Δa strength.

Conclusions: Table 2 compares the observed magnitude ranges, defined as the difference between the lowest and highest value over the rotational period, for Δa, Strömgren *uvby*, and Johnson *V*, respectively. In general, it can be concluded that absolute differences for Δa are much smaller than those for the other filters. This can be easily understood by the definition of Δa as measuring the flux at 5200Å and at the adjacent spectral regions (Sect. 3). Therefore, it is the difference of the amplitudes of the light variations at 5200Å and the mean of the adjacent regions. The effect of rotation is therefore minimized.

The total sample of investigated stars includes 17 CP2 objects (Table 1). This means that one star equals to 5.9% of the total sample. Since CP2 stars are normally slowly rotating in comparison to their apparent normal type counterparts (North, 1984), the rotational periods (1 to 130 d) of the sample represent the whole group very well. Also the magnetic field strengths seem to cover the range of the majority of CP2 stars (Bychkov et al., 2009). If one assumes the detection limit of +15 mmag which can be even easily achieved for much fainter Galactic open cluster members (Sect. 17.2.1), four stars (HD 30849, HD 38823, HD 72968, and HD 116458) would have been missed at certain phases of their rotational periods. The lower limits for these objects are +12, +6, +12, and +14 mmag, respectively, which is very close, but still below, the detection limit. One also have to emphasize that most targets with already very high absolute Δa values were chosen in order to study the possible effects of rotation.

The mean variation for all 17 stars is $\pm 8(5)$ mmag with a median value of ± 7 mmag, respectively. Taking these values and looking at the absolute values in Table 1, one can assume that the probability of a non-detection of an individual Galactic CP star due to a single observation at the minimum phase is about 10%. This probability can be significantly lowered by conduction individual observations during several nights and taking averaging values. The situation in the LMC on the basis of the available photometry is different (Sect. 19.5). There it was concluded that the rotationally modulated variability of the studied mCP candidates in the V and I bands is very weak (if present at all). It is very likely that the true amplitudes are much lower than 10 mmag, which is the limit derived from the time series analysis of the brightest mCP candidate. If one keeps in mind that the amplitude in Δa is much less than in the V and I bands, the probability of a non-detection is even significantly lower than in the Milky Way.

5.2 Be/shell stars - The case of Pleione

Maitzen & Pavlovski (1987a) presented Δa measurements of the Be/shell star HR 1180 (Pleione) within the Pleiades. It is a B8Vpe star known to exhibit prominent long-term photometric and spectroscopic variations with a period of about 35 yr (Tanaka et al., 2007). In the recent 70 years, Pleione showed twice a cyclic change between a Be-shell phase and a Be phase. For example, it entered a Be-shell phase in 1972, initiated by the appearance of a wide and shallow absorption of the Ca II K line. Then, the star developed many shell absorption lines in its spectrum. At the same time, the star had shown a decrease in its brightness. Pleione is known to be a speckle and a single-lined spectroscopic binary with an orbital period of 218 d as well as a large eccentricity of $\epsilon > 0.7$ (Nemravová et al., 2010). However, the astrophysical characteristics of the shell(s) and the possible circumstellar disk are still a matter of debate.

Measurements of Δa are clearly correlated with the shell and emission phase of Pleione (Pavlovski & Maitzen, 1989). A value of +36 mmag was obtained during the shell phase in December 1984 and then dropped almost to zero in the emission phase. The question arises how often a Be star in its shell phase mimics a classical CP star. This has obviously an influence on the incidence of via Δa detected CP stars in stellar aggregates, for example. Of course, all bona-fide photometrically detected CP candidates should be confirmed via spectroscopy. However, for more distant Galactic open clusters (Sect. 17.5) and the LMC (Sect. 19.4), the CP candidates are quite faint, even for 8 meter class telescopes for which observing time is, in addition, quite rare.

To investigate this issue, Pavlovski & Maitzen (1989) measured several times 40 bright and well known Be/shell stars via Δa photometry. From their sample, only two stars (5%) show a significant positive Δa value: HD 174638 (+36 mmag, β Lyr) and HD 193237 (+25 mmag, P Cyg). Both objects are quite outstanding and well investigated. Such objects could be easily sorted out via CMDs, for example, and should not be confused with CP stars, at all. Although HD 193237 is classified as supernova, there are strong indications that this object might have undergone a superoutburst typical of luminous blue variables (Chu et al., 2004). This shows the potential of Δa photometry to detect such objects which is especially important for studying star clusters of the Milky Way and extragalactic systems.

Finally, Pavlovski & Maitzen (1989) concluded from their statistical analysis that the contamination of CP2/CP4 stars by Be stars with positive Δa values is practically negligible, especially if one considers the colour (or $T_{\rm eff}$) range where CP2/CP4 stars are expected. Furthermore, they found that the behaviour of Pleione during its shell phase seems to be a rare phenomenon among Be/shell stars.

5.3 Variable star detection via Δa photometry

The detection of variable members of open clusters is very important since these objects have fairly well known astrophysical parameters, such as luminosity and effective temperature. Several theories (e.g. pulsational and evolutionary models) can be tested with these variable stars.

In the literature, a huge amount of papers dedicated to the search for new variable stars in open clusters can be found. In general, two different

Table 3: Open clusters observed at ESO and UTSO in 1995 (upper panel) as well as CASLEO in 1998 and 2001 (lower panel). The ages ($\log t$) and distance moduli ($V_0 - M_V$) were taken from the literature. The limit of apparent variability (Limit) is according to Sect. 5.3.1. The errors in the final digits of the corresponding quantity are given in parentheses.

Designation		N_S	N_V	N_F	N_N	JD (start)	Δt [d]	Limit [mag]	$\log t$	$V_0 - M_V$
NGC 2439	C0738−315	115	3	18	4	2449816.51736	7.995	0.022	7.30	13.00(10)
NGC 2489	C0754−299	53	1	13	4	2449818.57083	34.937	0.012	8.45	10.80(10)
NGC 2567	C0816−304	34	–	17	4	2449818.61111	44.868	0.012	8.43	11.10(10)
NGC 2658	C0841−324	84	1	12	3	2449817.56597	18.026	0.018	8.50	12.90(15)
Melotte 105	C1117−632	122	–	15	2	2449816.61111	3.072	0.010	7.77	11.30(10)
NGC 3960	C1148−554	32	–	17	3	2449828.60625	59.865	0.014	8.88	11.10(20)
NGC 5281	C1343−626	16	1	55	6	2449816.78125	48.010	0.008	7.04	10.60(15)
NGC 6134	C1624−490	82	2	32	6	2449821.75069	18.003	0.010	8.84	9.80(15)
NGC 6192	C1636−432	64	–	38	6	2449822.79583	44.949	0.006	7.95	11.15(20)
NGC 6208	C1645−537	15	–	12	2	2449880.69236	2.010	0.020	9.00	10.00(15)
NGC 6396	C1734−349	48	2	18	5	2449821.88958	31.939	0.014	7.40	10.60(15)
NGC 6451	C1747−302	41	2	12	2	2449883.78125	0.988	0.010	8.30	11.65(20)
NGC 6611	C1816−138	45	2	42	5	2449849.79306	38.989	0.016	6.48	11.65(10)
NGC 6705	C1848−063	275	1	43	5	2449822.86250	59.911	0.014	8.40	11.65(20)
NGC 6756	C1906+046	33	3	38	5	2449823.90069	53.944	0.008	8.11	12.60(15)
NGC 3114	C1001−598	181	7	50	4	2451138.82296	2.979	0.022	8.48	9.60(15)
Collinder 272	C1327−610	45	2	22	1	2452144.48472	0.020	0.008	7.11	11.85(15)
Pismis 20	C1511−588	178	2	80	2	2452143.52872	1.045	0.022	6.70	12.55(20)
NGC 6204	C1642−469	268	3	55	1	2452143.63456	0.067	0.020	8.30	10.40(25)
Lyngå 14	C1651−452	60	3	70	1	2452144.58135	0.087	0.008	6.00	12.05(15)

N_S: number of investigated stars; N_V: number of variable objects; N_F: number of frames; N_N: number of nights; Δt: time base of the observations

kinds of surveys are conducted: 1) the search for special types of variables (Viskum et al., 1997) or 2) selected open clusters are searched for all kinds of variable objects (Kafka & Honeycutt, 2003).

The search for new variable stars in Galactic open clusters is a serendipitous result from CCD Δa photometry (Sect. 17.3). The observations span widely different time intervals (0.02 to 60 d) yielding different possibilities for detecting the whole set of variations. It has to be emphasized that these observations are not optimized for the detection of variable stars but are able to find even very low amplitude variables (the typical detection limit reached is between 6 and 22 mmag). It is therefore a very important by-product or spin-off from Δa photometry.

5.3.1 Temporal analysis and results

The temporal analysis of the used photometric data is especially sophisticated since the overall time bases range from 0.020 to 60 days with 12 to 80 data points per individual cluster (Table 3). The smallest time resolution is about one minute with typically six frames within 30 minutes. As a consequence of the points discussed in the following, no individual light

Table 4: Photometric data for the Johnson UBV system taken from WEBDA for the identified variable stars; the mean reddening values for the open clusters are according to the literature. The lower panel includes the four variable objects which are most probably not members of the corresponding cluster. The objects are numbered according to WEBDA or to the internal numbering system (marked with asterisks).

Name	No.	$E(B-V)$	V	$B-V$	$U-B$
NGC 2439	128	0.37	14.72	0.33	0.22
	239		14.84	0.63	0.41
	741		15.91	1.61	
NGC 2489	28	0.40	15.46	0.93	0.21
NGC 2658	56*	0.44	16.73	0.64	
NGC 3114	14*	0.07	15.51	0.75	
	143*		15.83	0.64	
	193*		15.70	0.74	
	233*		16.29	0.70	
	272*		16.17	0.49	
	273*		15.86	0.96	
	274*		10.19	0.13	
Cr 272	1287	0.45	18.43	1.27	
	1297		14.21	0.57	
NGC 5281	1435	0.26	13.47	0.40	
Pismis 20	28	1.25	16.04	1.17	
NGC 6134	42	0.36	12.40	0.56	0.49
	662		15.14	0.50	
NGC 6204	92*	0.45	16.17	1.49	
	150*		14.80	1.10	
	360*		14.99	1.02	
Lyngå 14	115	1.48	14.99	1.38	
	150		14.78	1.21	
NGC 6396	1	0.97	9.79	1.82	−0.06
NGC 6451	199	0.70	14.06	0.62	
	716		14.46	0.71	
NGC 6611	198	0.85	13.21	0.60	−0.06
	343	1.16	11.72	0.87	−0.17
NGC 6705	770	0.43	13.72	0.50	0.42
NGC 6756	21	0.70	14.44	1.11	
	24		14.50	1.11	
Pismis 20	27	≈ 0	15.43	0.03	
Lyngå 14	101	≈ 0	14.39	0.62	
NGC 6396	20	high	10.62	2.44	2.63
NGC 6756	40	≈ 0	15.01	−0.01	

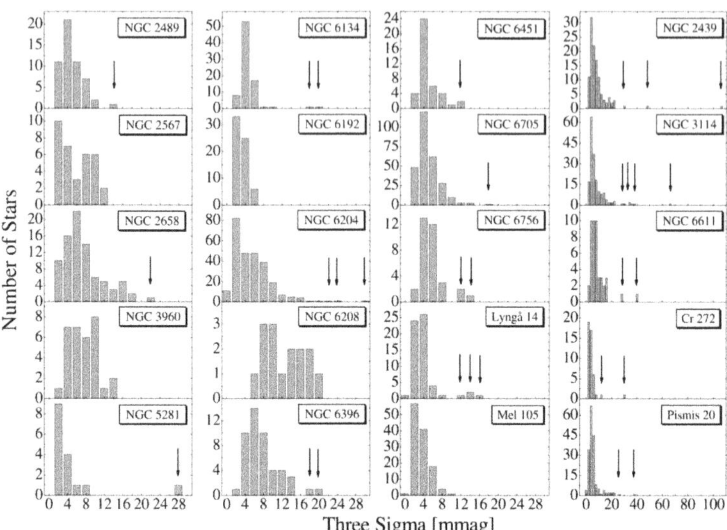

Figure 3: Histograms of the three sigma standard deviations of all mean photometric values for the programme clusters; arrows indicate the bona-fide variable stars detected.

curves are presented.

A classical time series analysis such as a Fourier technique (Paunzen & Vanmunster, 2016) cannot be performed since it is not optimized for sparse data sets with widely different time bases. The following approach was used to get a statistically solid limit for variability.

Since only a limited amount of data is available, all observations were added and analysed together. For each frame, a "standard mean magnitude" was calculated as the weighted mean photometric value (the weights are the measurement errors according to the PSF reduction technique) of all objects (variable and non-variable). The mean atmospheric extinction within the corresponding 700 Å decreases slightly with λ and may vary during the time of the observations (Schuster & Parraro, 2001). Furthermore, the quantum efficiency of CCD detectors increases towards the red region. Therefore, one is confronted with different zero points for the different standard mean magnitudes. The light curve of an "overall standard star" was used as comparison in the further analysis.

As a final step, differential light curves of each individual object in comparison to the "overall standard star" were generated. For all differential

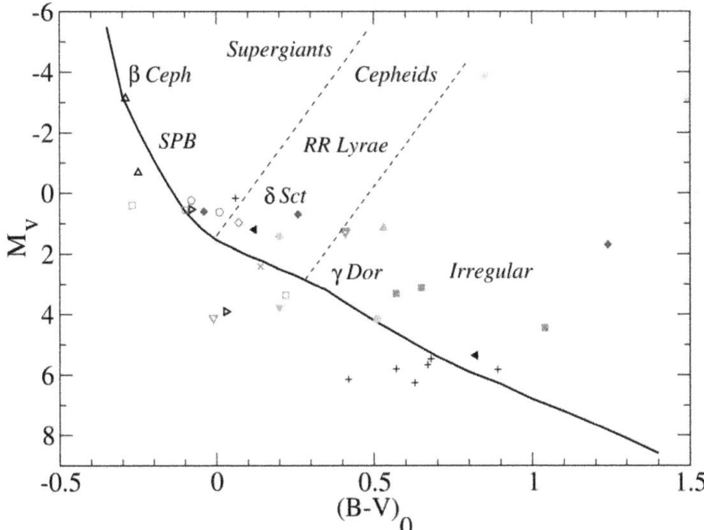

Figure 4: The HRD of the variable objects with the parameters listed in Table 4. The ZAMS was taken from Schmidt-Kaler (1982). The areas of different variable groups are indicated.

light curves, a mean magnitude and its standard deviation were calculated. An object was defined as variable if

- its standard deviation from the mean exceeds nine times the overall standard deviation of the cluster
- at least three data points exceed three times its standard deviation.

The first term is only the formulation of the statistical significance whereas the second guarantees that bad measurements or possible misidentifications do not affect the conclusions. The mean values are between 6 and 22 mmag (Table 3). As a test, a phase dispersion minimization analysis (Stellingwerf, 1978) was performed yielding the same statistical significance of variability. But the very unfavourable spectral window of the data sets prevents a definite conclusion about the true periods.

Figure 3 shows the histograms of the three sigma standard deviations of all mean photometric values of the observed cluster stars. The plotted standard deviations were normalized according to the errors of the photon noise. The influence of the number of data points on the detection limit is clearly visible (e.g. NGC 5281). No correlation was found between the amount of detected variable objects and the value of the detection limit.

5.3.2 Individual variables and Conclusions

Table 4 lists the 35 bona-fide variable objects found in 15 open clusters. The photometric data for the Johnson UBV system were taken from WEBDA[1]. Figure 4 shows the HRD of these objects with the parameters listed in Table 4. The ZAMS is taken from Schmidt-Kaler (1982). One is able to conclude from Table 4 and Fig. 4 that the following objects are *not* members of the corresponding open clusters because with the given $(B-V)$ colour and reddening, they would lie significantly below the ZAMS (Fig. 4): Pismis 20 # 27, Lyngå 14 # 101 and NGC 6756 # 40. However, with a reddening close to zero, they are very close to the ZAMS indicating that these objects are foreground stars. NGC 6396 # 20 seems to be a highly reddened background star. All other objects are probably members of the corresponding clusters; although no further membership information was found in the literature. One has to emphasize that the errors of $(B - V)$ are of the order of 0.05 to 0.3 mag depending on the type of measurement technique (photographic, photoelectric or CCD).

The literature was searched if variable objects have been confirmed for the investigated clusters. Rasmussen et al. (2002), for example, identified the variable stars in NGC 6134 (Table 4) and concluded that these are indeed known δ Scuti type objects with the possibility of a γ Doradus nature. Since the detection limit of variability for NGC 6134 is defined as 10 mmag, one can be confident that the employed method is valid for the given data sets.

Another important point is the type of variability. In Fig. 4, the position of known variable star groups have been indicated. Most of the objects seem to lie within the classical instability strip. Another large group is located within the area of the irregular variables (e.g. T Tauri objects). However, for an unambiguous conclusion about the true nature of the detected variability, follow-up observations, especially spectroscopy, are needed.

None of the objects with a peculiar Δa value shows evidence of variability which is in line with the conclusions listed in Section 5.1. This is an independent support that rotational induced variability of CP stars does not significantly influences the detection probability of Δa photometry.

The results of this investigation are another impressive example of how Δa photometric observations produce very important spin-off contributions for a completely different research field.

---

[1]http://webda.physics.muni.cz

6 Chemically Peculiar stars of the upper main sequence

The Δa photometric was designed to detect objects which exhibit a 5200Å flux depression, namely the chemically peculiar (CP) stars (Sect. 2). However, later it turned out that also several other objects of all spectral types and luminosity classes have peculiar features in this spectral region. Here, an overview of the different CP stars, including the group of λ Bootis stars are given.

The CP stars of the upper MS have been targets for astrophysical studies since the discovery of these objects by the American astronomer Antonia Maury (1897). Most of this early research was devoted to the detection of peculiar features in their spectra and photometric behaviour. The main characteristics of the classical CP stars are: peculiar and often variable line strengths, quadrature of line variability with radial velocity changes, photometric variability with the same periodicity and coincidence of extrema. Slow rotation was inferred from the sharpness of spectral lines. Overabundances of several orders of magnitude compared to the Sun were derived for Silicon, Chromium, Strontium, and Europium, and for other heavy elements.

Babcock (1947) discovered a global dipolar magnetic field in the star 78 Virginis followed by a catalogue of similar stars (Babcock, 1958) in which also the variability of the field strength in many CP stars – including even a reversal of magnetic polarity – was discovered. Stibbs (1950) introduced the Oblique Rotator concept of slowly rotating stars with non-coincidence of the magnetic and rotational axes. This model reproduces variability and reversals of the magnetic field strength. Due to the chemical abundance concentrations at the magnetic poles also spectral and the related photometric variabilities are easily understood, as well as radial velocity variations of the appearing and receding patches on the stellar surface.

Preston (1974) divided the CP stars into the following groups:

- CP1: Am/Fm stars without a strong global magnetic field; weak lines of Ca II and Sc II, otherwise strong overabundances;
- CP2: "classical" CP stars with strong magnetic fields, also known as the magnetic CP or mCP stars;
- CP3: HgMn stars, basically non-magnetic;

- CP4: He-weak stars, some of these objects show a detectable magnetic field.

The incidence of CP stars among all MS objects of the same spectral types can reach up to 15% (Netopil, 2013). Now, some more detail are given for these groups.

CP1 stars: The Am/Fm stars (CP1) are preferably found within close binary systems. The main characteristics of this group are the lack of magnetic fields, the apparent underabundance of calcium and scandium compared to the Sun, overabundances of Fe-peak elements, and very low rotational velocities. Almost all CP1 stars seem to be rather evolved with ages above 400 Myr (Künzli & North, 1998).

The observed abundance pattern is explained by the diffusion of elements together with the disappearance of the outer convection zone associated with the helium ionization because of gravitional settling of helium (Michaud et al., 1983). They predict a cut-off rotational velocity for such objects ($\approx 90\,\mathrm{km\,s^{-1}}$), above which meridional circulation leads to a mixing in the stellar atmosphere.

CP2 stars: This group of chemically peculiar stars was already described by Maury (1897). The main characteristics of the classical CP2 stars are: peculiar and often variable line strengths, quadrature of line variability with radial velocity changes, photometric variability with the same periodicity, and coincidence of extrema. Slow rotation was inferred from the sharpness of spectral lines. Overabundances of several orders of magnitude compared to the Sun were derived for heavy elements such as Silicon, Chromium, Strontium, and Europium.

The strong global magnetic fields exhibit variability of the field strength including even a reversal of magnetic polarity leading the Oblique Rotator concept of slowly rotating stars with non-coincidence of the magnetic and rotational axes. This model produces variability and reversals of the magnetic field strength similar to a lighthouse. Due to the chemical abundance concentrations at the magnetic poles spectral and the related photometric variabilities are also easily understood, as are radial velocity variations of the appearing and receding patches on the stellar surface (Deutsch, 1970).

Kodaira (1969) was the first to notice broad-band flux depressions at 4100, 5200 and 6300Å during his investigation of the CP2 star HD 221568. Photometrically, the main depressions (4100 and 5200Å) of another CP2 star were later also found by Maitzen & Moffat (1972) when they investigated the spectrum-variable HD 125248. These features were later also found to a weaker extend for other CP stars.

CP3 stars: The HgMn (CP3) stars are generally non-magnetic, slow rotating, B-type stars with certain chemical abundance anomalies, i.e. large excesses of (up to five orders of magnitudes) of Mercury and Manganese, but underabundances of Helium and Aluminium. Strong isotopic anomalies were detected for the chemical elements Calcium, Platinum, and Mercury, with patterns changing from one star to the next (Cowley et al., 2010). More than two thirds of the HgMn stars are known to belong to spectroscopic binaries (Hubrig & Mathys, 1995). It seems that the majority of slowly rotating late B-type stars formed in binary systems with certain orbital parameters become CP3 stars. There are several mechanism which play a major role in understanding these extreme peculiarities: radiatively driven diffusion, mass loss, mixing, light induced drift, and possible weak magnetic fields. However, there is no satisfactory model which explains the abundance pattern, yet (Adelman et al., 2003).

CP4 stars: As defined by Preston (1974), the CP4 stars comprise helium weak, B-type objects which often have strong and sharp lines of Gallium, Phosphorus, Titanium, Chromium, and Strontium. They have strong magnetic fields (as the CP2 group) which produce elemental surface inhomogeneities together with photometric variations. Several objects also show emission in the optical spectral range and signs of mass loss (Wahlgren & Hubrig, 2004).

λ **Bootis stars**: This small group comprises non-magnetic, late B- to early F-type, Population I, luminosity class V stars with apparently solar abundances of the light elements and moderate to strong underabundances of Fe-peak elements (Paunzen et al., 2002a). Only a maximum of about 2% of all objects in the relevant spectral domain are believed to be λ Bootis type stars. Several members of the group exhibit a strong infrared excess, and a disk (Paunzen et al., 2003b).

To explain the peculiar chemical abundances, Venn & Lambert (1990) suggested they are caused by selective accretion of circumstellar material. One of the principal features of that hypothesis is that the observed abundance anomalies are restricted to the stellar surface. On the basis of this hypothesis Kamp & Paunzen (2002) developed models which describe the interaction of the star with its local interstellar and/or circumstellar environment, whereby different degrees of underabundance are produced by different amounts of accreted material relative to the photospheric mass. The fact that the fraction of λ Bootis stars on the MS is so small would then be a consequence of the low probability of a star-cloud interaction within a limited parameter space. For example, the effects of meridional circulation dissolves any accretion pattern a few million years after the accretion has stopped.

7 Other star groups interesting for the Δa photometric system

Here, an overview of star groups other than the CP ones which are (possible) targets for Δa photometry is given.

7.1 Blue Stragglers

Originally, Blue Stragglers (BS) were discovered as members of open and globular clusters which are hotter than the turn-off point of the corresponding cluster (Mermilliod, 1982; Simunovic & Puzia, 2016).

Some main theories with several "flavours" were developed in order to explain the origin of BS (Stryker, 1993; Santucci et al., 2015):

- They are in a post-MS evolutionary state which coincides with the original MS
- They are normal stars formed long after the members of open and globular clusters
- They have accreted enough interstellar material from nearby stars with large mass loss rates or from a passing cloud to have moved up the MS
- They have acquired mass by partial or complete mass transfer from a binary companion
- They are the result of stellar collisions and coalescence
- They are undergoing quasi-homogeneous evolution in which substantial mixing which supplies the core with hydrogen beyond the normal time of exhaustion

None of these theories was proven or rejected by observations, yet. Probably, the phenomenon is caused not only by one of these theories but by a mixture of several ones.

Ahumada & Lapasset (2007) presented a catalogue of 1887 BS candidates in 427 open clusters. The extensive list, their characteristics, and statistical analysis of BS in Globular clusters is reviewed by Knigge (2015).

The situation about the definition of Field Blue Stragglers (FBS) is less satisfying because it is not possible to define a "classical turn-off point" for the field population.

Baade's original concept of a stellar "population" in the Milky Way was based on a disk (I) and a halo (II) population. The concept later evolved into a scheme with subdivisions. In one subdivision, the old metal-weak Population II was divided into halo (extreme) and intermediate Population II (ip II) objects. The latter was comprised of objects that had a velocity dispersion, a chemical composition, and a concentration towards the Galactic plane that was intermediate between those of the halo and the disk populations. The ip II population was later proposed in order to explain and fit both the global and the local Galactic kinematics. Robin & Crézé (1986) estimated that about 1.5 % of all stars in the solar neighbourhood are ip II objects.

However, Strömgren (1964) and Blauuw (1965) had already pointed out that, owing to a steady transition between the populations and also to insufficient precision in their ages, no clear distinction between the groups could be made. Strömgren (1966) therefore proposed using the $uvby\beta$ photometric system, as it would be more illuminating. The δm_1 index, for instance, was used to distinguish solar from underabundant objects in the spectral-type domain between early F and mid G.

Olsen (1979) addressed stars with $+0.175 < [m_1] < +0.215$ mag and $[c_1] < +0.935$ mag not being ip II as FBS. He was left with a very high percentage of stars according to the normal field objects (5 %). Carney & Peterson (1981) used the following criteria to find candidates of this group:

- $(B-V)_0 < +0.36$ mag (reddest turn-off color of globular clusters)
- Close to the MS in order to exclude FHB stars
- High proper motion and/or high radial velocity

Finally, they found only three candidates satisfying all criteria. This might already indicate the small percentage of true FBS which is not in line with the result of Olsen (1979).

Andrievsky et al. (1995, 1996) derived abundances for 18 stars of the list from Olsen (1980). They found mildly underabundances and rather low radial velocities for their sample. The mean value of the iron abundances was derived as $[Fe/H] = -0.31(13)$ dex whereas carbon is slightly enhanced $[C/Fe] = +0.32$ dex. The stellar parameters and the location in the HRD supported rather old ages for these stars (about 4 Gyr).

It has been shown by spectroscopic and Δa observations that up to 60% of the BS in open clusters show peculiarities of some kind. However, the absolute percentage is quite diverging between different references (Abt,

1985; Schönberner et al., 2001). In Sect. 17.4, the observations and their interpretations are described in more detail.

7.2 Herbig Ae/Be stars

Herbig (1960) proposed a new group of Pre-Main-Sequence (PMS) stars as higher mass counterparts of the T Tauri stars (Grankin, 2016) later known as Herbig Ae/Be stars. These intermediate mass ($2 - 8M_\odot$) stars are believed to be still in their radiative phase of contraction onto the MS. He defined the following criteria for a Herbig Ae/Be star:

- Spectral class earlier than F0
- Strong emission lines, particular those of the Balmer lines
- A location within the boundaries of a molecular cloud
- The star illuminates a nebula

It is obvious that the last two criteria ensure the PMS nature of these stars since the remnants of the protostellar cloud is still detectable. On this base, Finkenzeller & Mundt (1984) found nearly 60 stars fulfilling the classical criteria. Additional observations in the UV and IR region revealed further characteristics (Malfait et al., 1998) which lead to two new criteria added to the classical ones:

- Broad-lined, luminosity class V, IV and III, B- to F-type stars which show or have shown in the past emission-lines
- The presence of an IR excess due to circumstellar dust

The first criterion should allow to distinguish Herbig Ae/Be stars from evolved stars such as post-AGB stars which have strong IR excesses due to stellar winds.

Hipparcos photometry (van den Ancker et al., 1998) revealed variations for almost all known Herbig Ae/Be stars on the time scales of hours to weeks with highly variable amplitudes. Later on, also stellar magnetic fields and surface spots as for the CP2/4 stars (Sect. 6) were detected (Järvinen et al., 2016).

Miroshnichenko et al. (1999) presented classification resolution spectroscopic observations of 9 Herbig Ae/Be candidate stars. The found highly variable Fe II and Ti II emission lines in the region between 4900Å and

5500Å. Those emission lines influence all three "classical" (g_1g_2y) filters. However, they seem to be more concentrated and stronger in the g_2 filter which causes their detection within the Δa photometric system.

7.3 Be/shell stars

These stars are defined as non-supergiant B-type which have shown hydrogen emission in their spectra at least once. Phenomenologically, shell stars are stars with strongly rotationally broadened photospheric lines and additional narrow absorption lines. Some of the latter appear roughly at the centre of the photospheric instance of the same atomic transition while those of low excitation do not have a photospheric counterpart (Cowley et al., 2015).

Studies of Be stars in each of the accessible wavelength domains have revealed unique information: UV studies have emphasized the duality between Balmer emission line and the star's radiative wind flux. X-ray analysis have drawn a distinction between the high energy flux emitted from interacting binaries and most isolated Be stars (Torrejón et al., 2013).

Due to an equatorial disk produced by stellar winds, emission arises quite regularly. This disk is believed to be formed from the material ejected from the fast rotating central star. In addition, photometric variability on different time scales is a common phenomenon caused by the formation of shock waves within those disks. But also non-radial pulsation and variability due to rotation are observed (Porter & Rivinius, 2003).

The negative Δa values found (Sect. 11.9) are probably caused by emission of Fe and Mg lines in the spectral region from 5167 to 5197 Å (Hanuschik, 1987), which fall exactly within the g_2 filter and its bandwidth (Table 9).

7.4 B[e] stars

The B[e] phenomenon as defined by Conti (1976) is a designation for hot B-type stars whose spectra show forbidden and permitted emission lines of neutral and singly ionised atoms and strong infrared excess. These observed properties indicate very extended circumstellar matter and were discovered in only a small percentage of known B-type stars. The B[e] phenomenon nowadays incorporates not only B-type stars with forbidden emission lines in their optical spectrum, but it also includes stars from the pre-main sequence to compact planetary nebulae and supergiants (Lamers et al., 1998; Kraus et al., 2015).

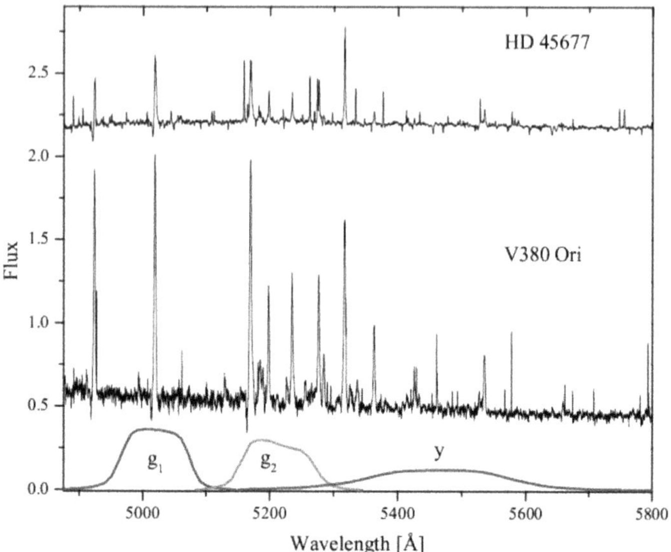

Figure 5: The spectra of the B[e] stars HD 45677 and V380 Ori from the ELODIE library together with the g_1, g_2, and g_3/y filters. Because the g_1 and g_3/y filters are also affected by emission lines, the usage of colours based on broad band filters is preferable.

This group attracted increasing attention in the last decade. The reasons for that are the development of new and more efficient ground-based instruments, the availability of space-based observations, the extension to other than the optical wavelength regions, and the development of new models. However, the number of objects showing this phenomenon is still very limited. The detection can be done either by time-consuming spectroscopy (search for the forbidden emission lines) or photometry (IR-excess, for example). Both methods have their severe limitations due to the high demand of observing time and the ambiguity of the different photometric diagrams (Dunstall et al., 2012).

The spectra of two B[e] stars, HD 45677 and V380 Ori, were taken from the ELODIE (Prugniel & Soubiran, 2001). Figure 5 shows the spectra of these B[e] stars together with the g_1, g_2, and g_3/y filters. It is clearly visible that the g_2 filter measures several forbidden Fe II emission lines. Notice how the absolute flux varies for these two objects. However, g_1 is significantly affected by the $H\beta$ line which is in emission whereas within g_3/y there are again some Fe II emission lines. It seems therefore preferable to use colours based on broad band filters for which this effect is less pronounced.

From Fig. 5 it is obvious that B[e] stars can be detected via observations in the g_2 filter. This will be especially important for the identification of these stars in the Magellanic clouds (Sect. 19).

7.5 Evolved and Population II stars

In Globular clusters several different highly evolved star groups can be found: red giant and sub-giant branches (RGBs and SGBs) as well as blue horizontal-branch objects (BHBs and HB). These groups are defined and characterized as

- (B)HB: horizontal-branch stars in Galactic globular clusters are old, low-initial mass ($0.7 - 0.9\,M_\odot$) stars currently burning helium in their core. Owing to their complexity, some aspects of the formation and internal structure of HB stars have been a matter of debate for decades (Catelan, 2009). For instance, the blue extension of the HB varies among clusters, a fact partly associated with the metallicity but not entirely explained by it. A number of second parameters in addition to metallicity, such as stellar rotation, cluster concentration, presence of super-oxygen-poor stars, cluster mass, environment of formation, and the cluster age have been suggested to provide an explanation for this behaviour, but none has proven completely adequate in describing the complex observational picture.

- SGB: The subgiant branch is a stage in the evolution of low to intermediate mass stars after the MS until the RGB. As the fraction of hydrogen remaining in the core of a MS star decreases, the core temperature increases and so the rate of fusion increases. This causes stars to evolve slowly to high luminosities as they age and broadens the MS band in the HRD. Once a MS star ceases to fuse hydrogen in its core, the core begins to gravitationally collapse (Goudfrooij et al., 2015). This causes it to increase in temperature and hydrogen fuses in a shell outside the core, which provides more energy than core hydrogen burning. Low- and intermediate-mass stars expand and cool down until at about 5000 K they begin to increase in luminosity in the RGB. The shape and time scale of the SGB phase varies for stars of different masses, due to differences in the internal structure. The SGB lasts for about 12 Myr for a star of $3\,M_\odot$, and only 1 Myr for a star of $6\,M_\odot$. Therefore, this evolutionary phase is so short that the chance of observing objects in this phase is very small. This led to the so-called Hertzsprung gap in the CMD of intermediate-age stellar systems (Adamczak & Lambert, 2014).

- RGB: The red-giant branch is the portion of the giant branch before helium ignition. It is a stage of stellar evolution that follows the MS and SGB for low- to intermediate-mass stars. RGB stars have an inert He-core surrounded by a shell of hydrogen fusing via the CNO cycle (Cohen et al., 2015). Along the RGB, the He-core mass does monotonically increase due to the conversion of H into He produced by the H-burning shell. Due to the release of gravitational energy by slow contraction, it tends to heat up. This favours the persistence of an isothermal core temperature profile. At the same time, with increasing densities, the neutrino production in the central regions of the core becomes progressively more efficient. This favours cooling of the innermost layers of the core. This leads to a feedback of the H-burning shell with the temperature and density of the He-core which then alters the stellar surface luminosity. The overall metallicity, i.e. C, N, and O, controls this mechanism which is the so-called He-core mass-luminosity relation for RGB stars. The RGB phase ends when the off-centre maximum temperature reaches the value to ignite He-burning via triple-α captures and is called He-Flash.

Mainly on the basis of Hubble Space Telescope (HST) observations, it was found that several globular clusters have not only at least two MS, which are explained by a different helium content (Piotto et al., 2007), but also simultaneously different HBs, RGBs, and SGBs (Grundahl et al., 1998; Piotto et al., 2012).

There are different "jumps" in the BHB distribution as well as peculiar HB extensions, like the blue-hook (Brown et al., 2001). However, the cause of these phenomena is still not clear, but it is probable connected to the complex star formation history of the individual clusters (Valcarce et al., 2012).

Gratton et al. (2012) gave an excellent overview of all types of chemical peculiarities for members of globular clusters. Not only variations in lighter elements (Li, C, N, O, Na, Al, and Mg), but also of CN and CH were detected. Possible correlations and relations of the individual abundances for different elements were widely analysed in the context of stellar formation and evolution. In general terms, one has to distinguish, as for Population I type objects, between intrinsic and photospheric peculiarities.

An intriguing phenomenon was first reported by Behr et al. (1999) who found large deviations in element abundances from the expected cluster metallicity for BHB stars in the globular cluster NGC 6205. For example, Fe was found to be a factor of three enhanced compared to the solar value, or about 100 times the mean cluster iron abundance. Such atmospheric effects are well known and studied for the classical CP stars of the upper MS (Sect. 6). Furthermore, in Sect. 14 it was concluded that Fe is mainly responsible for producing a positive Δa index, but also Cr and Si are contributors at least for lower effective temperatures. Later on, Khalack et al. (2010) found clear evidence for vertical stratification of Fe in atmospheres of BHBs. In addition, it was reported that BHB stars with $T_{\rm eff} \geq 11\,500$ K have lower rotational velocities than their cooler counterparts (Behr, 2003b) suggesting that the atmospheres of such stars are stable enough to have atomic diffusion. The models by Quievy et al. (2009) show that He sinks in stars with low rotational velocities, which leads to the disappearance of the superficial He convection zone. This then opens the door for atomic diffusion to play a role. The atmospheric models of Hui-Bon-Hoa et al. (2000) showed that the observed photometric jumps and gaps for hot BHB stars can be explained by elemental diffusion in their atmosphere. These models calculate self-consistently the structure of the atmosphere while taking into account the stratification predicted by diffusion (assuming equilibrium). They confirm that vertical stratification of the elements can strongly modify the structure of the atmospheres of BHB stars (LeBlanc et al., 2009). Such structural changes for the atmosphere lead to the photometric anomalies discussed above. However, the search for strong magnetic fields in HB stars was not successful (Elkin, 1995).

In overall it can be concluded that these slow-rotating BHB stars are the Population II type counterparts of the classical CP stars (Sect. 6).

Inspecting the spectral region around 5200Å, where the g_2 filter is centred, for RGB, G- and K-type MS stars (Gray & Corbally, 2009), strong Mg I (5167Å, 5173Å, and 5183Å) and MgH features are found (Sect. 8.2). This was also the reason why the Vilnius Z filter has been centred in this region (Sect. 4). Those features strongly vary with the evolutionary status, i.e. for dwarfs and giants. Johnson et al. (2005) derived magnesium abundances for more than 100 RGB stars in each of the Galactic globular clusters NGC 5272 and NGC 6205. They found a wide spread of magnesium abundances within the individual aggregates (about 0.7 dex) and some significant outliers. So it might be possible to detect strongly deviating elemental abundances of Mg and luminosity effects with Δa photometry.

8 An inspection of the wavelength region from 4900 to 5700Å

The original narrow-band Δa photometric system with its three filters covers approximately the wavelength region from 4900 to 5700Å. Let us recall here the characteristics of the three filters (Maitzen & Vogt, 1983): g_1 ($\lambda_c = 5010$ Å, FWHM $= 130$ Å), g_2 (5215, 130) and g_3/y (5485, 230).

These three filters were always used in two ways/combinations (Sect. 3): a versus $(g_1 - y)$ and y versus $(g_1 - y)$ where $a = g_2 - \frac{g_1+y}{2}$. So, in principle, any other colour combination, also with those from other photometric filter systems, can be applied to study any possible astronomical objects. Normally, all individual $g_1, g_2, g_3/y$ observations, especially those performed via the CCD technology (Sects. 17.3, 18, and 19), are archived and available for any follow-up investigation.

In the following, a quantitative overview of the corresponding wavelength region for stars, quasars, and galaxies is given. For this, spectral libraries are used to show the characteristics of several different peculiar and "normal" type objects. Furthermore, a summary of the current knowledge about the 5200Å flux depression is included.

8.1 The 5200Å flux depression

As stated before (Sect. 2), the Δa photometric system was primarily intended to investigate the 5200Å flux depression found for CP stars. Within this work, a spectrophotometric investigation of this region (Sect. 9) and an approach to derive synthetic Δa values (Sect. 14) are presented. In these Sections, many details about the characteristics of this flux depression are listed. Furthermore, several correlations of the depth with astrophysical parameters and the elemental compositions for different star groups are listed. The interested reader is referred to these Sections. Here, a short summary of the 5200Å flux depression of CP stars is given.

Kodaira (1969) was the first who noticed significant flux depressions at 4200 Å ($2.35\,\mu^{-1}$), 5300 Å ($1.90\,\mu^{-1}$), and 6300 Å ($1.55\,\mu^{-1}$) in the spectrum of HD 221568 (Fig. 6). The latter is a magnetic CP2 star with an extremely large amplitude (0.2 mag in $B-V$) rotational induced variability and a period of about 160 d (Stępień & Muthsam, 1987). The flux distribution varies strongly between the red and blue phases. Unfortunately, no Δa value for it was measured, yet. For some time, the central wavelength of this depression was believed to be at 5300 Å (Adelman, 1975) because it is a very broad feature investigated with rather low resolution spectropho-

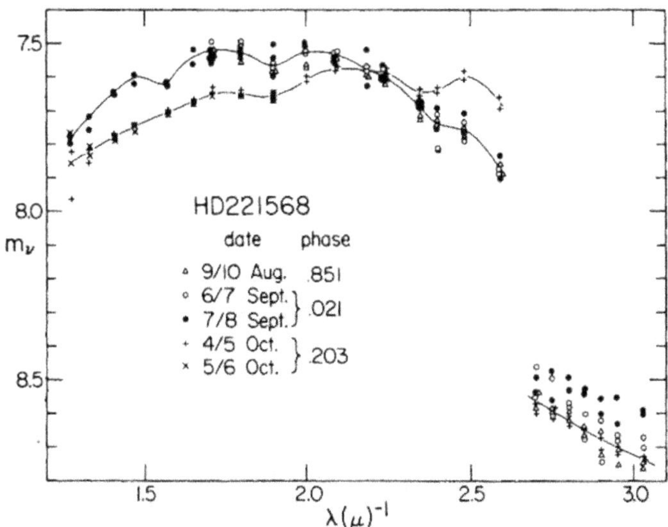

Figure 6: The first notification of significant flux depressions at 4200 Å ($2.35\,\mu^{-1}$), 5300 Å ($1.90\,\mu^{-1}$), and 6300 Å ($1.55\,\mu^{-1}$) in the spectrum of HD 221568 by Kodaira (1969).

tometry only. Maitzen (1976a) was the first who presented evidence that the centre of the two bluer depressions are actually very close to 4100 and 5200 Å, respectively. Jamar (1977, 1978) detected and investigated similar features in the ultraviolet region at 1400 Å, 1750 Å, and 2750 Å showing that flux depressions are not a phenomenon restricted to the optical.

Several attempts have been made to explain the origin of this feature. Adelman & Wolken (1976) investigated bound-free discontinuities, Jamar et al. (1978) proposed autoionisation transitions of Si II, whereas enhanced line absorption was discussed by Maitzen (1976a) and Maitzen & Muthsam (1980). The latter presented a comparison of synthesised flux distributions and observed spectrophotometry. The first attempt in this direction was made by Leckrone et al. (1974). From their synthetic spectra, Maitzen & Muthsam (1980) recovered a narrow and deep feature at about 5175 Å and a broad component centred at about 5275 Å. They were not able to reproduce the flux depression for effective temperatures higher than 8000 K. More recently, Adelman et al. (1995) have used ATLAS9 model atmospheres (Kurucz, 1993) with enhanced metallicity (i.e. a solar element abundance distribution where all elements heavier than He had been scaled by +0.5 or +1.0 dex) and concluded that at least part of the 5200 Å feature in magnetic CP stars may be due to differential line blanketing. In a follow-up work, Adelman & Rayle (2000) have extended this study to a larger group of stars and found that solar composition models may successfully predict the flux distribution of normal and many CP3 (HgMn) stars, while they fail to do so for a number of CP2 stars, i.e. the group of stars showing the largest flux depression in the 5200 Å region and thus the largest magnitudes in Δa.

Khan & Shulyak (2007) found that the majority of the tested chemical elements (within the limits of abundance values considered) produce less than 1% variations in the model atmosphere temperature profile. These elements are He, C, N, O, Mg (deficient), Ca, Sr, Eu, and Hg. Several elements produced moderate changes on the model atmosphere structure (1 – 3%) and the energy distribution. These elements were Mn, Ni, and Mg (enhanced). The group of elements which produced large changes in the model atmosphere structure (more than 3%) and energy distribution consisted of Si, Cr, Fe and scaled abundance patterns.

They found that Fe is the principal contributor into the 5200Å depression for the range of $T_{\rm eff}$ between 20000 and 8000 K, while Cr and Si are important primarily for low $T_{\rm eff}$.

8.2 Features of Stars across the HRD

First, the method of spectral classification was developed and used in the second half of the nineteenth century. Thanks to the invention of photographic plates and the long standing knowledge of prism, for example, it became a powerful tool quite parallel to photometry. Even a brief history of its development would fill many pages. The interested reader is referred to the book by Hearnshaw (2014) who gave an excellent overview about two centuries of astronomical spectroscopy.

A whole zoo of different systems have been developed from which only the Harvard and Yerkes systems survived. The latter is associated with the work of Morgan et al. (1943) who used a prismatic spectrograph covering the region from 3900 to 4900 Å with a dispersion of 115 Å mm^{-1} at Hγ. During the decades, new developments (for example, grating spectrographs and the CCD technology) made it necessary to check and revise the original systems. Again, the interested reader is referred to two excellent books by Jaschek & Jaschek (1990) and Gray & Corbally (2009) who listed all available information about the spectral classification in great depth.

It has to be emphasized that the wavelength region for the classical spectral classification is between 3900 and 4900 Å, respectively. The filters of the Δa photometric system are actually redward of this region. This

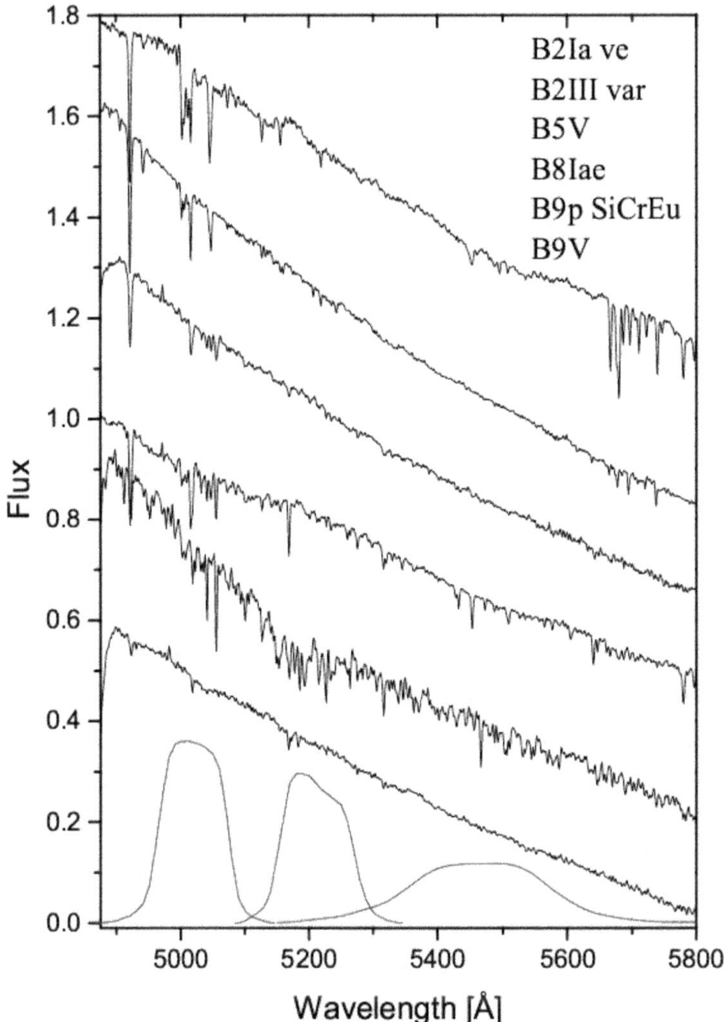

Figure 7: Spectra for selected B-type stars from the MILES library together with the $g_1, g_2, g_3/y$ filter curves.

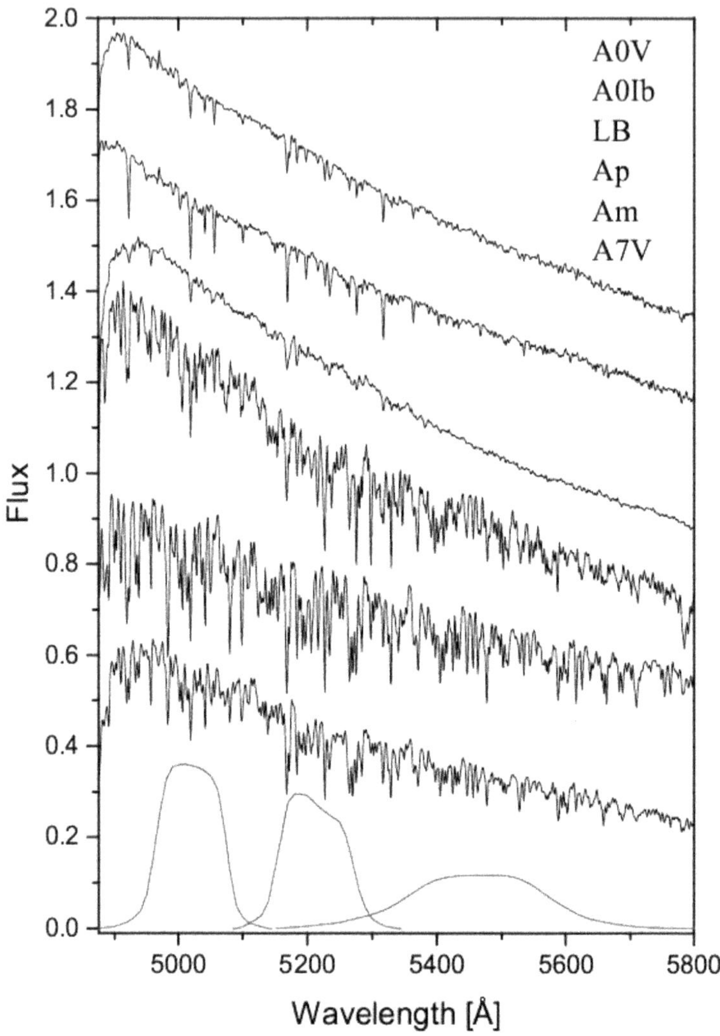

Figure 8: Spectra for selected A-type stars from the MILES library together with the $g_1, g_2, g_3/y$ filter curves.

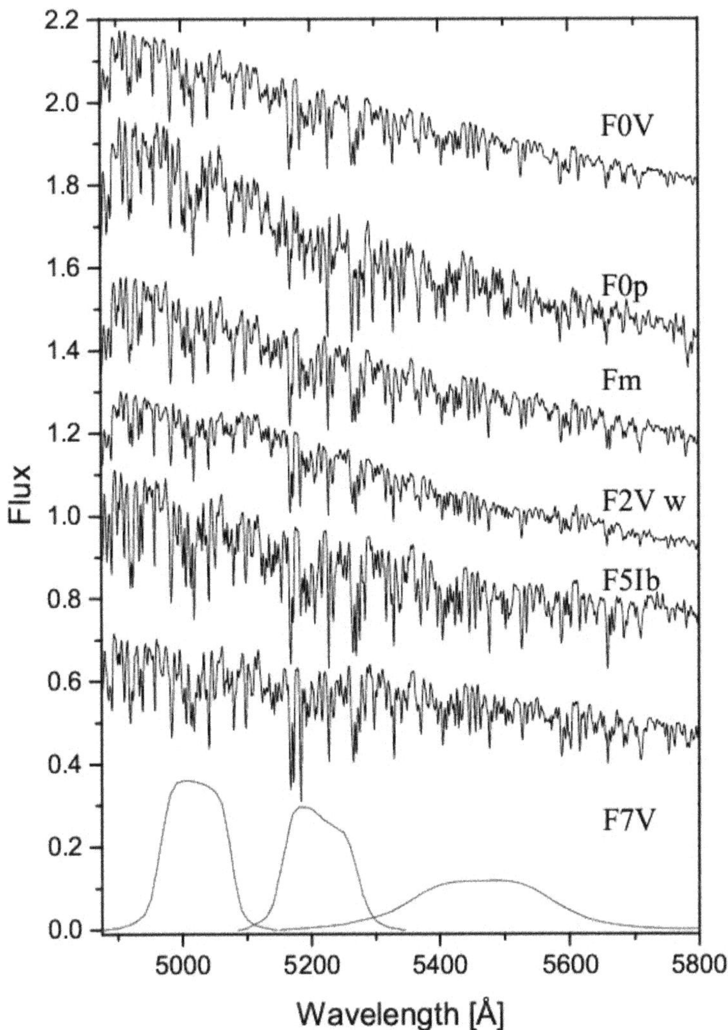

Figure 9: Spectra for selected F-type stars from the MILES library together with the $g_1, g_2, g_3/y$ filter curves.

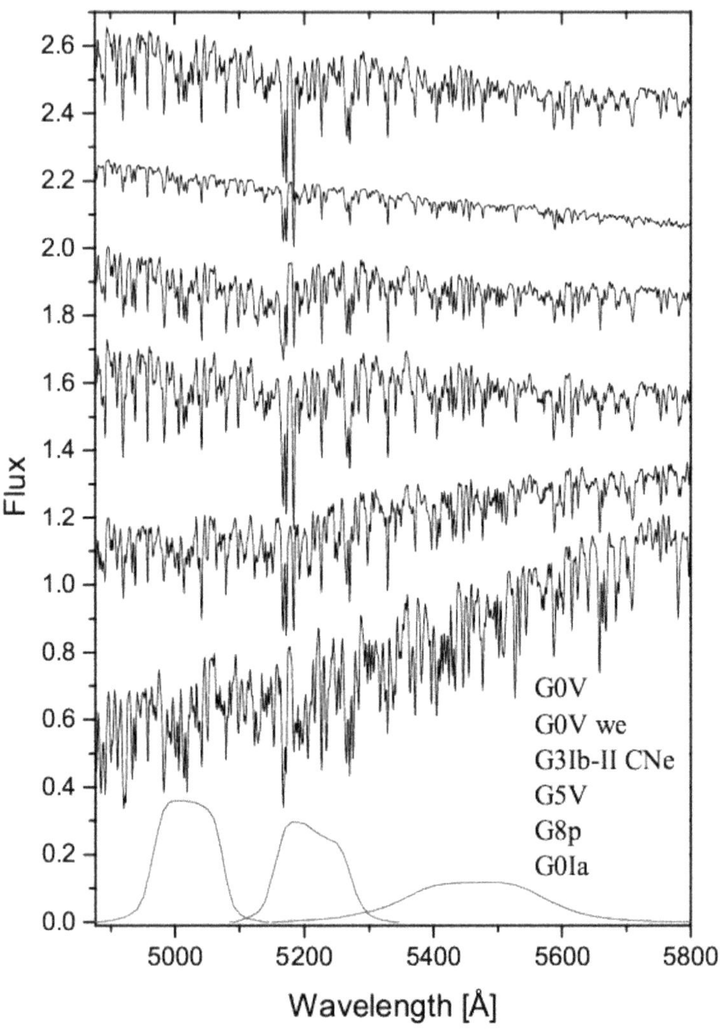

Figure 10: Spectra for selected G-type stars from the MILES library together with the $g_1, g_2, g_3/y$ filter curves.

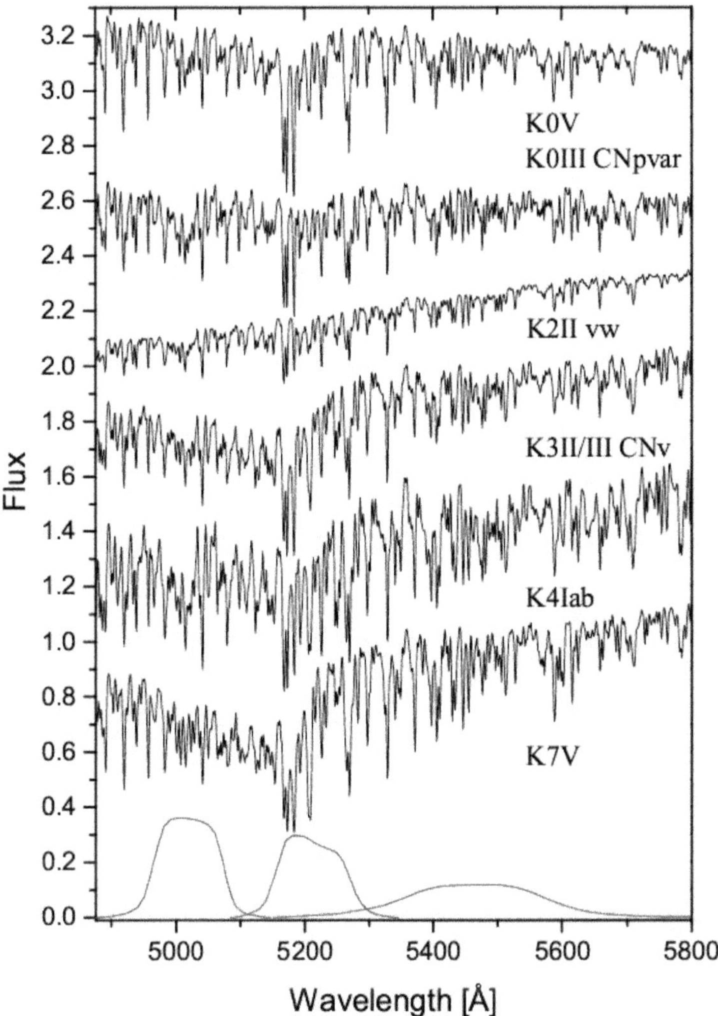

Figure 11: Spectra for selected K-type stars from the MILES library together with the $g_1, g_2, g_3/y$ filter curves.

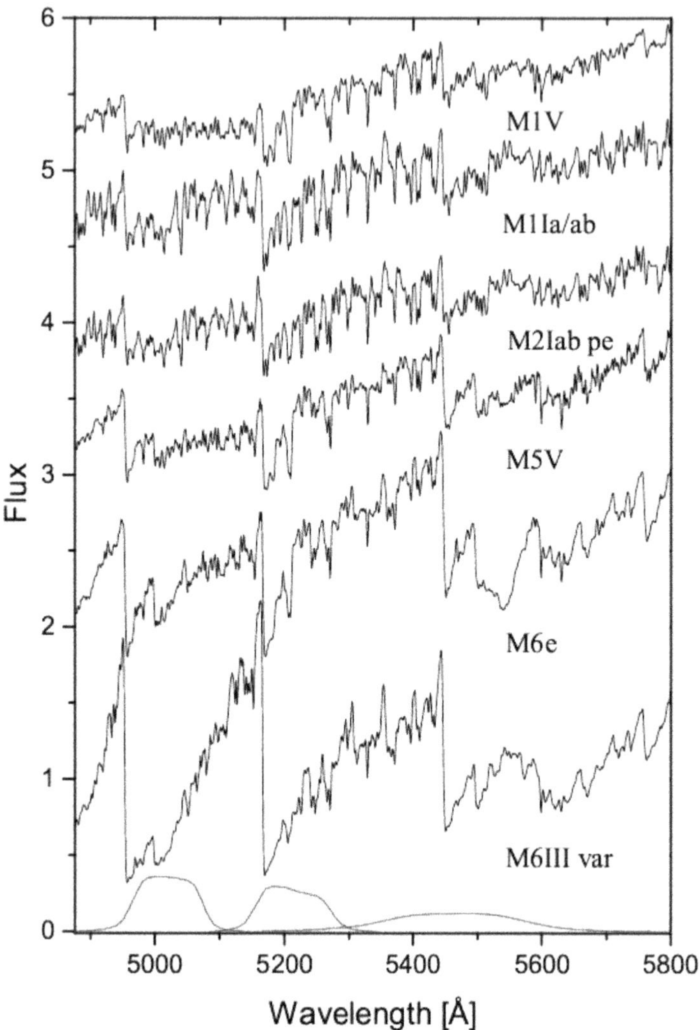

Figure 12: Spectra for selected M-type stars from the MILES library together with the $g_1, g_2, g_3/y$ filter curves.

means that the spectral features there are not used for the classification process. The only exceptions are molecular absorption bands for very cool type stars. Those features are, for example, C_2 (5164Å), CN (5560Å), MgH (5211Å), TiO (5167, 5448, and 5497Å), and ZrO (5305 and 5379Å).

In the following, the corresponding spectral region is shown for B- to M-type stars of all types. This is just a short excerpt of many peculiar objects over the whole HRD and by far not complete. It should show only a cross section of already investigated and potential targets for the Δa photometric system.

The spectra were taken from the MILES library which consists of 985 stars spanning a large range in atmospheric parameters and spectral types (Falcón-Barroso et al., 2011). The spectra were obtained at the 2.5 m Isaac Newton Telescope at the Observatorio del Roque de los Muchachos and cover the range from 3525 to 7500Å with an originally estimated spectral resolution of 2.3Å (full-width half maximum, FWHM). A homogenized compilation of the stellar atmospheric parameters ($T_{\rm eff}$, $\log g$, and [Fe/H]) for the stars of this library was presented in Cenarro et al. (2007). The stellar library is also the basis for our empirical models for stellar population synthesis (Vazdekis et al., 2010). Both, the stellar library and models can be easily accessed and manipulated through this website: http://miles.iac.es. All spectra are calibrated in absolute flux unites and can be directly compared.

Figures 7 to 12 show the corresponding spectra for B- to M-type stars. The individual spectra were only shifted in the flux unit to present them in a convenient way. Each figure also includes the $g_1, g_2, g_3/y$ filter curves as a guidance of the covered corresponding spectral regions. In the following, the different spectral regions are discussed in more detail.

B-type stars (Fig. 7): There are strong variations, especially for the non-MS objects in g_1, visible. Also strong lines for the B2Iave object in the red wing of g_3/y are evident. On the other hand, the flux depression at 5200Å is clearly visible for the B9p SiCrEu (CP2) star whereas no emission lines for any star are present.

A-type stars (Fig. 8): There is a significant increase of the metallic lines spectrum from A0 to A7 for the complete spectral region. Also the metallic lines of the A0Ib supergiant are stronger than for the A0V star, although their $v \sin i$ values are comparable. This probably also causes that some supergiants have positive Δa values. The metal-weakness of the λ Bootis and the strong overabundances of the Am (CP1) and Ap (CP2) stars are evident. Other than the Am star, the Ap object shows the 5200Å flux depression.

F-type stars (Fig. 9): All spectra are very similar in the overall metallic-line strength and the spectral flux distribution. Only the Fp object shows slightly stronger metal-lines compared to the corresponding F0V standard star. From the point of view of the Δa photometric system, the F-type stars are the least interesting ones to follow in forthcoming projects.

G-type stars (Fig. 10): The first remarkable feature is the change of the flux distribution for the G8p and G0Ia objects as the g_3/y magnitude becomes significant brighter. Such a behaviour should result in a blueing of the colour. The G0Vwe star and G3Ib-II CNe supergiant clearly show much weaker metallic-line spectra than the corresponding standard MS stars. These star groups, especially the weak-lined one, should be easily detectable via Δa photometry.

K-type stars (Fig. 11): For this spectral region, also a flux depression within the g_1 filter is visible. However, contrary to the CP stars, it is most significant for apparent normal type MS objects. As for the G-type objects, the weak-lined group seems to by quite outstanding among the sample. On the other hand, the K4Iab supergiant shows stronger lines as its MS counterparts. Again, this behaviour is very similar to the hotter supergiants.

M-type stars (Fig. 12): The objects up to a spectral type of M5 ($T_{\mathrm{eff}} \approx 3000\,\mathrm{K}$) have quite a homogeneous flux distribution over the complete spectral region. For cooler temperatures, the bandheads become very steep with a significant strong flux distribution. This characteristic might be used to search for M-type stars with large radial velocities. Such shifts should alter not only the colours but also the a index.

As a summary, it can be said that the additional capabilities of Δa photometric system are manifold over the complete spectral type range besides maybe for the F-type stars for which no exceptional outliers were found. Some guidelines for new projects and/or a re-analysis of archival data are listed in Section 22.

8.3 Features of Quasars

Other than stars, no dedicated observing programme to find and study quasars in the 5200Å has been conducted, yet. A quasar (also denoted as quasi-stellar object or QSO) is an active galactic nucleus of very high luminosity. It consists of a supermassive black hole surrounded by an orbiting accretion disk of gas. As gas in the accretion disk falls toward the black hole, energy is released in the form of electromagnetic radiation (Liu et al., 2016). Interesting enough, also optical/ultraviolet variability of these objects has been discovered to be correlated with other quasar properties,

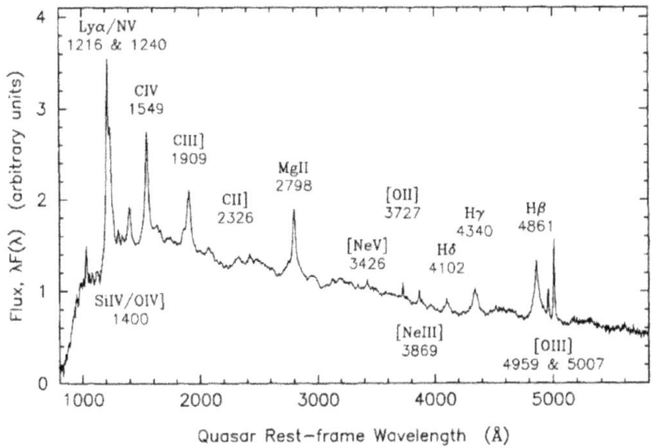

Figure 13: A very high signal-to-noise ratio composite spectrum of the rest-frame ultraviolet and optical region of a high luminosity quasar together with the identified features (Francis et al., 1991).

such as luminosity, black hole mass and rest-frame wavelength. To date, about 450 000 quasars over a very broad range of distances (corresponding to redshifts of up to 7) are catalogued. However, they are faint in the optical region with 3C 273 being the brightest Quasar known with $V = 12.0$ mag.

The most prominent emission type features of quasars can be found in the UV region as shown in Figure 13. The shown spectrum is a composite of 718 individual spectra of quasars published by Francis et al. (1991). The strongest features have equivalent widths of almost 60Å as listed in Table 5. The latter also includes the calculated redshifts z which shift the corresponding lines to the centre of the g_2 filter. The strongest features cover redshifts from 0.07 to 3.28, respectively.

Finding new Quasars can be done either by searching for the above listed features in the optical spectral region as done by the Sloan Digital Sky Survey (Pâris et al., 2017) or by analysing multi-band photometry (Brescia et al., 2013). Having precise photometry from the UV to the IR region, due to the redshift, the features affect different filters to a certain amount. This behaviour can be easily simulated. However, it is not straightforward to distinguish redshifted Quasars from other objects. Brescia et al. (2013) developed a machine learning method that can be used to cope with regression and classification problems on complex and massive data sets in order to identify Quasars from photometric catalogues. They used the GALEX, SDSS, UKIDSS, and WISE catalogues which give only

Table 5: The strongest identified features shown in Fig. 13 together with the equivalent width and the redshift z shifting them in the centre of the g_2 filter.

λ_{rest} [Å]	Species	Eqw [Å]	z	λ_{rest} [Å]	Species	Eqw [Å]	z
5007	[O III]	15	0.04	1549	C IV	37	2.36
4861	Hβ	58	0.07	1400	Si IV/O IV]	10	2.71
4340	Hδ	10	0.20	1240	N V	52	3.19
2798	Mg II	50	0.86	1216	Lyα	52	3.28
1909	C III]	22	1.72				

14 000 objects common in all four data bases. This is a surprisingly low number. The more different filters are available, the better this method works.

A dedicated project searching for new Quasars using photometric observations in the g_2 filter seems not efficient because already several deep all-sky survey are available. However, the observations in the Magellanic Clouds could be used to search for Quasars (Sect. 19). The data of the Magellanic Quasars Survey (Kozlowski et al., 2013, MSQ) are ideal to test the method using g_2 to identify Quasars. They spectroscopically confirmed 758 Quasars selected from the third phase of the Optical Gravitational Lensing Experiment based on their optical variability, mid-IR, and/or X-ray properties. About 50 Quasars are bright enough ($I < 18$ mag) to immediately study with the available g_2 data. After this pioneer study, the feasibility of further projects can be evaluated.

8.4 Features of Galaxies

Besides stars, star clusters, and the Magellanic Clouds, galaxies should not be lost out of sight. With the modern CCD technology and large telescopes, even distant galaxies could be observed within the Δa photometric system. Two interesting projects in this respect are the the Advanced Large Homogeneous Area Medium-Band Redshift Astronomical (ALHAMBRA, Moles et al., 2008) and Cosmic Evolution (COSMOS, Scoville et al., 2007) survey which are described in Section 21. The first project will observe a total area of four square degrees on the sky with the focus to study galaxies of all types. There is one filter centred at 5220Å which perfectly matches g_2 but with a broader FWHM of 310Å. The second project is designed to probe the formation and evolution of galaxies as a function of both cosmic

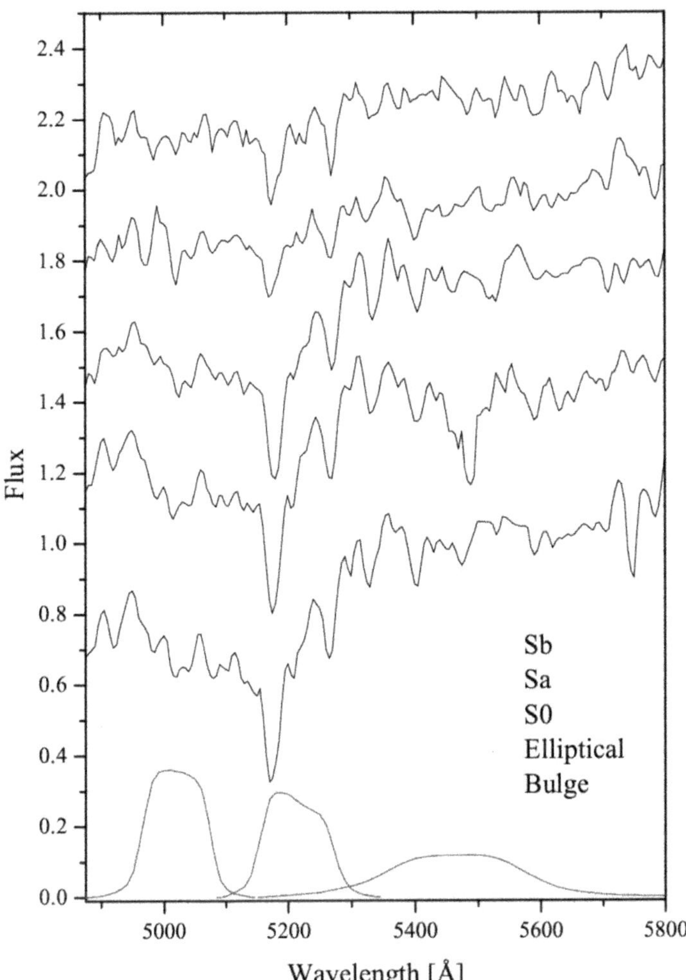

Figure 14: Spectral templates for bulge, elliptical, S0, Sa, and Sb galaxies (from bottom to top) from Calzetti et al. (1994) and Kinney et al. (1996) together with the $g_1, g_2, g_3/y$ filter curves.

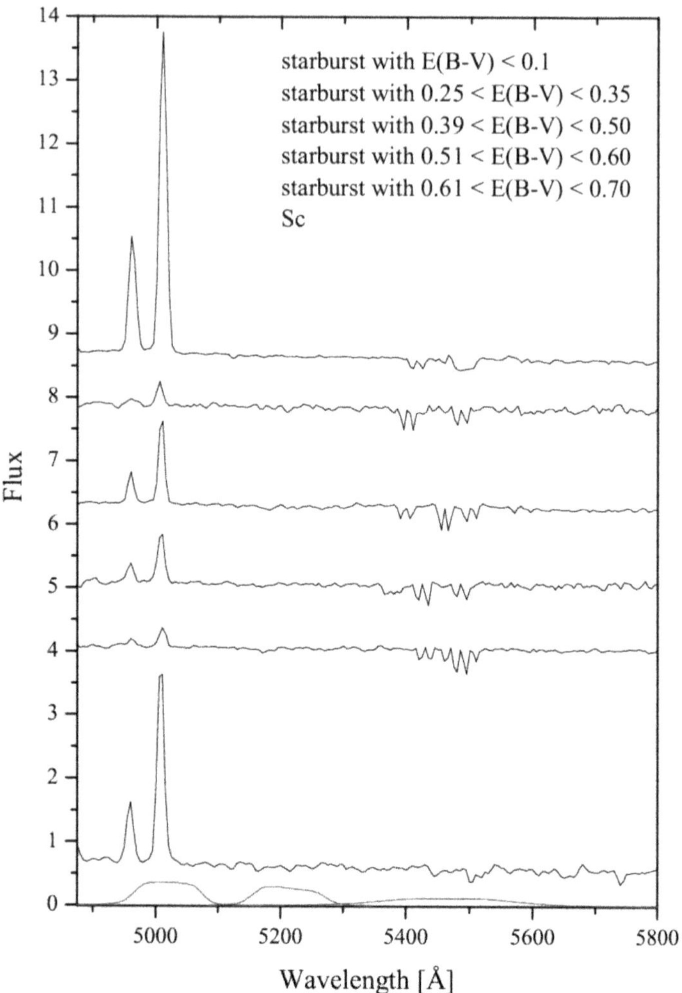

Figure 15: Spectral templates for Sc and starburst galaxies with different internal extinction values from Calzetti et al. (1994) and Kinney et al. (1996) together with the $g_1, g_2, g_3/y$ filter curves.

time (redshift) and the local galaxy environment. The survey covers a two square degree equatorial field with imaging by most of the major space-based telescopes and a number of large ground based telescopes, with many ongoing surveys. It includes the "Subaru Telescope Intermediate Band Filter" IA527 which is centred at 5259Å with a FWHM of 242Å.

Because of the availability of filters similar to g_2, the spectra of various galaxy types in this spectral region were investigated in more detail.

The spectral templates of the various galaxies types were taken from Calzetti et al. (1994) and Kinney et al. (1996). They have a two pixel resolution of 10Å and are calibrated in absolute fluxes. The spectral templates cover various galaxy morphological types from elliptical to late type spiral. Starburst templates for low, $E(B - V < 0.10$ mag, to high, $0.61 < E(B - V < 0.70$ mag, internal extinction are also available. Several of the starburst galaxies used in the construction of the starburst templates are classified as irregulars. Thus, although irregular galaxies are no explicitly covered, the starburst templates can be used to cover this morphological type. All details about how these templates were built and which individual galaxies were used can be found in Calzetti et al. (1994) and Kinney et al. (1996).

It has to be emphasized that none of the parameters such as different metallicities, dust/gas ratios, star forming episodes, and so on have been taking into account here. These parameters alter the spectra of galaxies to a certain amount (Möller et al., 1997).

Figures 14 and 15 show the templates for the galaxies divided in those with and without emission features, respectively. All conclusions drawn below are for a zero redshift. If the latter is taken into account as non-zero, the analysis of the capability of the Δa photometric system becomes much more complex. One example for such an analysis of the COSMOS survey can be found in Mobasher et al. (2007).

Bulge, elliptical, S0, Sa, and Sb galaxies (Fig. 14): The Sa and Sb spectral templates are almost indistinguishable from each other. The bulge, elliptical, and S0 show discriminative features in the g_3/y filter. Also the overall flux distributions are slightly different. However the Sa/Sb and bulge/elliptical/S0 galaxies can be clearly distinguished from each other by the g_2 filter.

Sc and starburst galaxies (Fig. 15): When comparing the spectral region of the g_2 with those of the g_1 and g_3/y filters, it is clear that the classical a index will not work for these galaxies. There are almost no changes in g_2 compared to the adjacent regions visible. The emission features [O III] at 4959Å and 5007Å (both sampled by the g_1 filter) are varying significantly for all galaxy types. There are also several strong diversify absorption features (mainly due to Fe and Mg) between 5380 and 5510Å (g_3/y). For these types of galaxies it seems best to use g_1 (preferable) or g_3/y instead of g_2 in order to construct an a_{Gal} index as $a_{\mathrm{Gal}} = g_1 - \frac{B+V}{2}$, for example.

For a zero redshift, the Δa photometric system is able to a-priori distinguish between Sa/Sb and bulge/elliptical/S0 galaxies on the basis of a classical a versus $(g_1 - y)$ diagram. However, this type of diagram malfunctions to distinguish between different types of Sc and starburst galaxies which have strong emission lines in the g_1 filter. Therefore, a discrimination between galaxies with and without emission features will not be possible. Taking into account a redshift, the conceptual formulation becomes quite complex. Depending on the amount of the redshift, different spectral features can be traced with the different filters. Such an analysis will require detailed observations of a significant amount of different galaxy types at a wide range of redshift values. Up to now, no such data sets are available.

9 Spectrophotometry of the 5200Å region for peculiar and normal type stars

Using spectrophotometry, Kodaira (1969) was the first to notice broadband flux depressions at 4100, 5200 and 6300Å during his investigation of the CP2 star HD 221568. Photometrically, the main depressions (4100 and 5200Å) of another CP2 star were later also found by Maitzen & Moffat (1972) when they investigated the spectrum-variable HD 125248.

A study was conducted to detect CP stars on the basis of spectrophotometric data for which three different methods were employed that use the flux depression at 5200Å. In addition to the original method, the a' system developed by Adelman (1979) and a newly modified method were investigated. Furthermore, the capability of these systems of finding new, previously undetected, peculiar stars including metal-weak and emission-type objects of the upper MS was analysed . Also, the low-mass regime was searched for the behaviour of the a indices in correlation with the effective temperature and luminosity.

9.1 Data and reduction

The source of spectrophotometric data is the stellar catalogue by Kharitonov et al. (1988). It contains data for 1159 bright stars of different spectral types in the wavelength range from 3200 to 7600Å, with a spectral resolution of 50Å. Each star was observed at least three times, on different

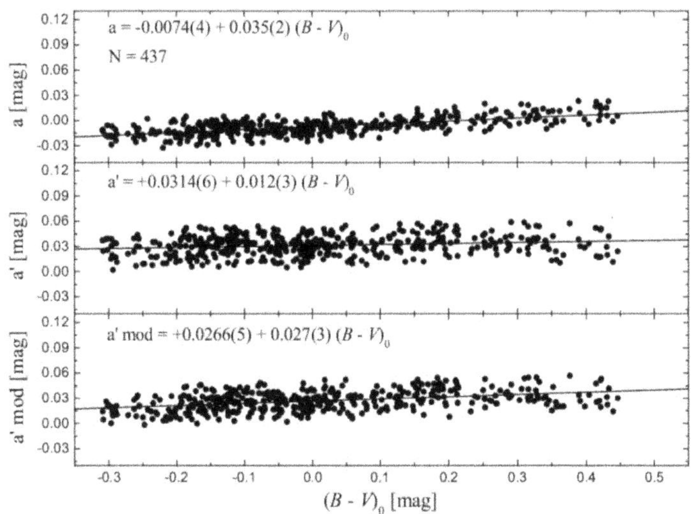

Figure 16: Regression of 437 normal stars with reference to the different a indices.

nights. The root mean square (rms hereafter) relative error of the measurements is between 2% and 4% per 50Å bandpass. Stars with only two-digit measurements in the wavelength region around 5200Å were not considered in the analysis because of the statistical inaccuracy of these measurements. Early-type supergiants were excluded because of their strong photometric and spectroscopic variability. The remaining data of 1067 stars were used to synthesize different peculiarity indices. The methods used to detect CP stars are designed to have magnitudes as input. Therefore the intensity and flux values given in the catalogue were converted into magnitudes.

One hast to be aware that there are more extensive corresponding catalogues with higher resolutions available in the literature, for example the one by Burnashev (1985). This author, for example, also used at that time unpublished data by Kharitonov et al. (1988), but interpolated the data to a resolution of 25Å. However, the here used catalogue is the basis of a list of spectrophotometric standards published by Glushneva et al. (1992). Therefore the quality of the data is beyond any doubt. In addition, the spectral resolution is similar to that used for the Gaia mission (Jordi et al., 2010).

Spectrophotometric observations of classical CP stars were published before, for example by Maitzen & Muthsam (1980) and Adelman et al.

(1989). However, all these analyses concentrated either on individual stars or on the comparison with results from synthetic spectra.

The goal of this analysis is not only to detect CP stars, but also to compare the findings with already existing Δa photometry and spectrophotometry. For this purpose, three different systems were employed (Sect. 4).

1. The Δa photometric system by Maitzen (1976a): For simplicity, the three filters were represented by Gaussian curves with a FWHM of 130Å and a transmission maximum of 100% each. The fact that these simulated filters, in contrast to the original definition of g_1, g_2 and y, all have the same transmission curves, probably does not significantly affect Δa, because its value is always the difference between a peculiar star and the corresponding normal star with the same colour index. To convolve the Gaussian curves with the measurements of the spectrophotometric catalogue, transmission values of 50Å (bins) of the simulated filters, corresponding to the measurements with 50Å spectral resolution, had to be determined. This was achieved by segmenting the Gaussian curves into 5Å bins, which were then numerically integrated. The percentage of transmission, relative to the maximum, could now be calculated by applying the mean value theorem. For the transmission values of the 50Å bins, to be multiplied with the measurements, simply ten of the 5Å bin transmission values were summed. This approach enables to adapt the central wavelengths of the filters in 5Å steps to figure out the optimal filter positions. The central wavelengths of the filters were optimised with respect to a low variance of a values for normal stars and secondly the highest possible a values of CP2 stars. The best compromise of filter positions found for the spectrophotometric data is 5020Å for g_1, 5215Å for g_2, and 5470Å for y. These filter positions are very close to those of the revised Δa system by Maitzen (1980b). The values of the simulated filters were calculated by multiplying the transmission values of the 50Å bins with the corresponding measurements of the catalogue and adding the results.

2. The $\Delta a'$ index by Adelman (1979): This index is described in more detail in Section 4.

3. The $\Delta a' mod$ index: Very early in the process of evaluating the spectrophotometric data the central wavelengths of the spectrophotometric data did not fit those of the $\Delta a'$ index very well. To calculate the a' index, the measurements closest to the defined wavelengths were

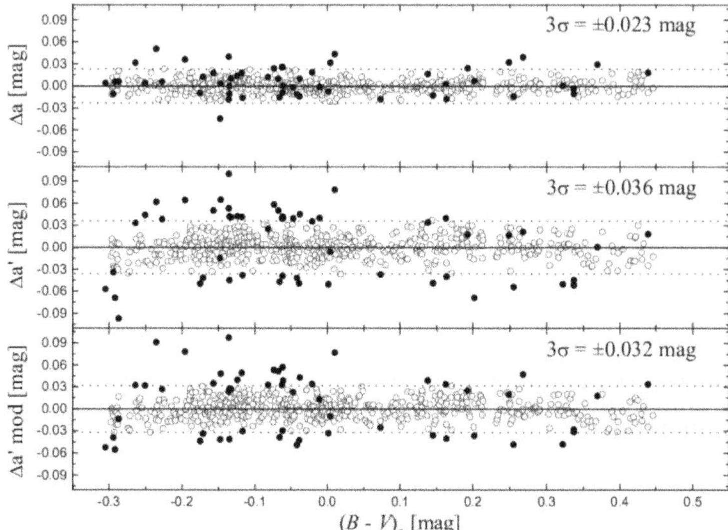

Figure 17: Peculiarity indices of 631 stars in the spectral-type range from O6 to F5. The filled circles denote stars that are detected as peculiar within at least one index.

used. This resulted in inaccurate results, because the central measurement was shifted from the deepest part of the depression to the red side of the spectra and in some stars the $H\beta$ line influenced the continuum value on the blue side. To address this problem two values next to each other were linearly interpolated to obtain central wavelengths closer to the predefined wavelengths of the $\Delta a'$ index, as well as to improve the error by a factor of $\sqrt{2}$. This new index is referred to as $\Delta a'mod$ and is defined as

$$a'mod = m_{5250} - [m_{4750} + 0.453 \cdot (m_{5850} - m_{4750})]. \qquad (11)$$

The calculation of the indices a' and the $a'mod$ was straightforward. They were obtained by simply applying the values of the corresponding measurements to the formula.

As next step, the indices had to be corrected with the a, a', and $a'mod$ obtained from non-peculiar stars of the same temperature. For this purpose, all stars hotter than F5 were selected to calculate a_0, a'_0 and a'_0mod.

According to Maitzen (1976a), the indices are expected to be very well correlated with the colour index $(B-V)_0$. To deredden the programme stars, use was made of photometric data in the Johnson, Geneva, and Strömgren systems, compiled from the General Catalogue of Photometric Data (GCPD, Mermilliod et al., 1997). For the first two systems, the well-known calibrations based on the X/Y parameters (Cramer, 1999) and Q-index (Gutierrez-Moreno, 1975), respectively, which are applicable for O/B type stars were applied. Objects with available Strömgren data were treated with the routines by Napiwotzki et al. (1993), which allow dereddening of cooler-type stars. For about 60 % of the targets it was possible to determine reddening values, transformed to $E(B-V)$ with the ratios summarised by Netopil et al. (2008). If several estimates for a particular object were available, a mean value was calculated. Since all objects are rather bright and therefore close-by, a strong reddening especially for cool-type stars is hardly expected. Therefore non-reddening was adopted for these when no Strömgren photometry was available, which is justified by the obtained overall mean $E(B-V) = 0.02(6)$ mag.

To find the normality lines, an iterative linear regression of a, a' and $a'mod$ versus $(B-V)_0$ was performed. In each iterative step, the outliers that lay more than 5σ from the normality line were discarded. After three iterations, the errors did not significantly decrease any more. In total, the normality lines is defined from 437 stars and a 3σ of 23, 36, and 32 mmag was obtained, respectively. Figure 16 shows the regression of the peculiarity indices versus $(B-V)_0$.

The three Δa indices were calculated by subtracting the corresponding a_0, a'_0 and a'_0mod from the a, a' and $a'mod$ values for the same $(B-V)_0$. Figure 16 shows the peculiarity indices versus $(B-V)_0$. All stars above the threshold of 3σ are expected to be chemically peculiar, whereas those below are emission-type and/or metal-weak objects. The region of $(B-V)_0$ in which CP2 stars are expected (hotter than spectral type F5) contains 631 stars (Fig. 17).

At $(B-V)_0$ of about 1.5 mag, corresponding to a spectral type M0, the indices a, a' and $a'mod$ lose their correlation with $(B-V)_0$ although they are nearly linearly correlated at bluer colour indices. This may indicate a dependency of a, a' and $a'mod$ on the luminosity class of the stars (see Fig.18).

9.2 Results

The findings of this investigation are summarized in Tables 6 and 7. The first table lists all well-known CP stars taken from the catalogue by Renson & Manfroid (2009). The second table includes apparently normal-type

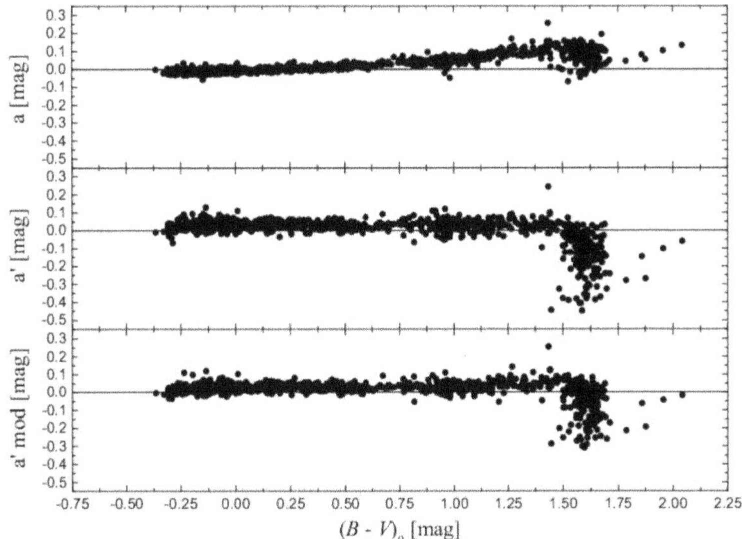

Figure 18: Different a indices of 1067 stars from the catalogue by Kharitonov et al. (1988) plotted versus $(B-V)_0$.

objects (according to spectral classification) hotter than spectral type F5 that have Δa values both higher and lower than the 3σ thresholds. The tables contain the HD number, spectral type, the Δa values of each applied method, and the CP class (Table 6 only) of the stars. For comparison with existing peculiarity measurements, they also include published Δa values (Sect. 10) and $\Delta(V1-G)$ (Sect. 4) values.

9.3 Known peculiar objects

First of all, the 55 CP stars listed in Table 6 that are included in the catalogue by Renson & Manfroid (2009) were analysed. Altogether, 19 stars were detected beyond the 3σ threshold by any of the systems. The modified system of Adelman yields 18 stars above 3σ for this sample. This is the highest ratio of spectrophotometric to otherwise identified peculiar stars of the three applied detection methods.

In Fig. 19, the observed $\Delta(V1-G)$ versus the synthetic $\Delta a'mod$ values for the sample is presented. The observed Δa values were not used because there are too few available. There is a clear correlation of the observed and synthesized values. The five outliers (HD 11415, HD 115735, HD 189849,

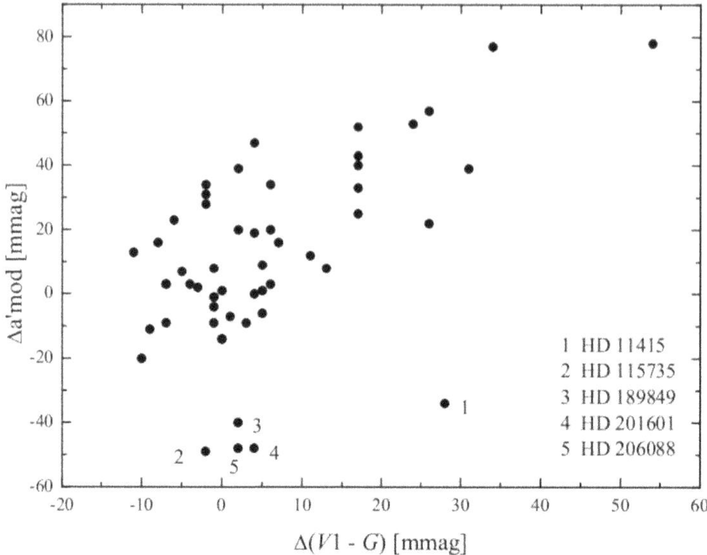

Figure 19: Observed $\Delta(V1 - G)$ versus synthetic $\Delta a'mod$ values (Δa and $\Delta a'$ behave very similar) for the CP star sample. The outliers are discussed in the text.

HD 201601, and HD 206088) are described below.

The CP1 (Am) stars have, with some rare exceptions, observed Δa values well below +10 mmag and are normally inconspicuous in the $\Delta(V1-G)$ index. However, four stars were detected, three above and one below, the 3σ threshold; these are

- *HD 29479*: a moderate overabundance of Fe-peak elements and [Ba/H] = +1.77 dex compared with that of the Sun was reported by Iliev et al. (2006)

- *HD 76756*: α Cancri is one of the prototype Am stars, with no other conspicuous features

- *HD 173648*: this is a hot Am star with overabundances of most Fe-peak elements, and considerable overabundances of Sr, Y, Zr, and Ba as well as some rare earths (Adelman et al., 1999)

- *HD 189849*: a weak magnetic field of about 250 G was detected on a 20σ level (Bychkov et al., 2009), so it might be misidentified.

It seems that some very peculiar CP1 stars are detectable via spectrophotometric observations and the Δa peculiar indices.

The magnetic CP2 stars are mainly detectable via Δa photometry, which is reflected in the results. Figure 20 shows the spectrophotometric data of standard star HD 102647 (A3 V) and HD 118022 (A2 Cr Eu Sr) for the whole spectral range (upper panel) and the region where the Δa system is situated (lower panel). The UV excess (Sokolov, 2006) and the 5200Å depression, both typical for CP stars, are clearly visible for HD 118022. All strong positive photometric detections are reproduced by the synthetic investigation. There are two objects, that have significant negative synthetic Δa values, but have statistically insignificant observational values:

- *HD 201601*: γ Equulei is a well-studied rapidly oscillating Ap star with a strong magnetic field and a very peculiar abundance pattern (Perraut et al., 2011).

- *HD 206088*: a very peculiar object, in a transition between CP1 and CP2 shows strong variability in the infrared (Catalano & Renson, 1994).

The spectrophotometry of both objects might be severely influenced by the unusual elemental peculiarity and the variability.

Among the apparent non-magnetic CP3 (HgMn) stars, no detection in any system was found. This is perfectly in line with the results from Section 11.5.

The sample includes only three well-known CP4 stars of which two (HD 11415 and HD 115735) show significant negative synthetic $\Delta a'$ and $\Delta a' mod$ values. Both objects are known to have strong emission lines according to Kohoutek & Wehmeyer (1999) which explains the result.

9.4 Apparent normal-type objects

Table 7 lists the apparently non-CP stars detected via synthetic photometry. This sample can be divided into emission-type and metal-weak objects as well as inconspicuous stars.

The hot-emission-type stars are defined as dwarfs that have shown hydrogen emission in their spectra at least once. Due to a developed equatorial disk that is produced by stellar winds, emission arises quite regularly. In addition, photometric variability on different time scales is a common phenomenon caused by the formation of shock waves within those disks. The phases of emission are replaced by shell and normal phases of the

Table 6: Synthetic (Δa, $\Delta a'$, and $\Delta a'mod$) and observed ($\Delta a\,obs$ and $\Delta(V1-G)$) peculiarity indices in mmags for well-established CP (flag "$*$") stars taken from Renson & Manfroid (2009). The synthetic photometric values for detected objects are given in boldface italics.

HD	SpType	Δa	$\Delta a'$	$\Delta a'mod$	$\Delta a\,obs$	$\Delta(V1-G)$	CP group
6116	A3–A9	+8	+20	+16	–	−8	1
20320	A2–A9	−6	−13	−14	−3	0	1
27045	A3–F3	−3	+1	−4	–	−1	1
27962	A1–A4	+13	+26	+31	–	−2	1
28355	A5–F1	−1	+2	+2	–	−3	1
29140	A3–A7	+12	+25	+23	–	−6	1
29479	A3–A9	+6	***+37***	+28	–	−2	1
40536	A4–F1	−3	+18	+16	+7	+7	1
41357	A4–F2	−4	+6	−1	–	−1	1
76756	A3–F1	+17	+34	***+39***	–	+2	1
116657	A1–A7	−6	+19	+25	+10	–	1
125337	A2–A7	−11	−8	−9	–	−7	1
141795	A2–A8	−1	−6	+3	–	−7	1
173648	A4–F1	+3	***+40***	***+34***	–	−2	1
173654	A2–A7	+3	+17	+13	–	−11	1
189849	A5–A9	−17	***−40***	***−40***	–	+2	1
197461	A7–F0 dD	−6	−10	−7	–	+1	1
198743	A3–F3	+8	+5	−9	–	−1	1
207098	A5–F4 dD	−4	+7	+1	–	+5	1
223461	A3–F0	+3	+7	+8	–	−1	1
10221	A0 Si Sr Cr	+12	+25	***+33***	–	+17	2
11502/3	A1 Si Cr Sr	+10	***+45***	***+43***	+39	+17	2
15089	A4 Sr	−1	−4	0	–	+4	2
19832	B8 Si	+14	***+43***	***+40***	+10	+17	2
26571	B8 Si	+2	−25	−20	+14	−10	2
32549	B9 Si Cr	+10	***+50***	***+52***	+25	+17	2
32650	B9 Si	+8	+12	+12	–	+11	2
34452	B9 Si	***+36***	***+65***	***+78***	–	+54	2
40312	A0 Si	+7	+30	+25	–	+17	2
68351	A0 Si Cr	***+24***	***+58***	***+53***	+24	+24	2
90569	A0 Sr Cr Si	***+26***	***+42***	***+57***	+36	+26	2
112185	A1 Cr Eu Mn	+18	+35	***+34***	–	+6	2
112413	A0 Eu Si Cr	+13	+19	+22	+40	+26	2
118022	A2 Cr Eu Sr	***+43***	***+78***	***+77***	+51	+34	2
120198	A0 Eu Cr Sr	+1	***+40***	***+39***	+38	+31	2
137909	A9 Sr Eu Cr	***+39***	+21	***+47***	+25	+4	2
148112	A0 Cr Eu	−6	+7	+8	–	+13	2

Table 6: continued.

HD	SpType	Δa	$\Delta a'$	$\Delta a' mod$	$\Delta a\,obs$	$\Delta(V1-G)$	CP group
170000	A0 Si	+7	+6	+6	–	–	2
183056	B9 Si	+13	+23	+20	–	+2	2
201601	A9 Sr Eu	−14	***−54***	***−48***	+10	+4	2
206088	A7–F3 Sr	+1	***−50***	***−48***	+7	+2	2
358	B9 Mn Hg	0	−7	−6	–	+5	3
23950	B9 Mn Hg Si	+3	−19	−14	–	0	3
33904	B9 Hg Mn	−1	+3	+3	–	−4	3
35497	B8 Cr Mn	+3	+4	+3	–	−7	3
77350	B9 Sr Cr Hg	−4	+10	+9	+1	+5	3
78316	B8 Mn Hg	0	+18	+20	+12	+6	3
106625	B8 Hg Mn	+3	−18	−11	–	−9	3
129174	B9 Mn Hg	−4	+4	+1	–	0	3
143807	A0 Mn Hg	−7	+13	+3	–	+6	3
145389	B9 Mn Hg	+15	+8	+19	–	+4	3
220933	A0 Hg Mn	−3	−2	+7	–	−5	3
11415	B3 He wk.	+13	***−41***	***−34***	–	+28	4
23408	B7 He wk. Mn	+13	−9	−9	+5	+3	4
115735	B9 He wk.	−11	***−42***	***−49***	+2	−2	4

same object, leading to a transition from negative to positive Δa values (Pavlovski & Maitzen, 1989). Among the metal-weak objects, the most prominent subgroup are the λ Bootis stars (Sect. 6). The Δa observations of the two groups are summarized in Section 11.

In the sample of inconspicuous objects, two β Cephei pulsators (HD 19374 and HD 29248) were detected with significantly high positive Δa values. These objects are early-type-B stars with light and radial velocity variations on time scales of several hours (Stankov & Handler, 2005). Since these objects are quite rare, fewer than 200 Galactic objects are known, it would be interesting to find out whether they can be detected via Δa photometry. Unfortunately, no observations in this respect have been performed so far. The remaining objects have two characteristics in common:

1. Except for one object (HD 22951), all of them are very fast rotators ($v \sin i \geq 150\,\mathrm{km\ s^{-1}}$) which is atypical for CP stars.

2. Almost all objects are in binary systems, which might distort the spectrophotometry.

For the following objects additional interesting characteristics were found in the literature:

Table 7: Synthetic (Δa, $\Delta a'$, and $\Delta a'mod$) and observed ($\Delta a\,obs$ and $\Delta(V1-G)$) peculiarity indices in mmags for apparent non-CP stars detected via synthetic photometry. The synthetic photometric values for detected objects are given in boldface italics. The upper panel lists well-known emission-type and metal-weak objects.

HD	SpType	Δa	$\Delta a'$	$\Delta a'mod$	$\Delta a\,obs$	$\Delta(V1-G)$
5394	B0.5 IVe	+17	***−37***	−21	+5	+1
6811	B7 Ve	−10	***−45***	***−41***	−5	+2
18552	B8 Ve	***+40***	***+100***	***+97***	–	−2
22192	B5 Ve	−9	***−49***	***−44***	–	−4
23016	B8V (e)	−16	***−38***	−30	–	−6
32537	F1 Vp MgII 4481Å weak	−10	***−51***	−31	–	−7
34078	O9.5 Ve, var.	+4	***−57***	***−52***	–	−10
35439	B1 Vpe	−11	−34	***−49***	+5	+3
67934	A0 Vnp MgII 4481Å weak	−1	***+40***	+13	–	−4
74873	kA0.5hA5mA0.5V λ Boo	−18	***−37***	−25	–	−15
111604	A5 Vp λ Boo	−11	−34	***−39***	–	−2
112014	A0 IIsp Mg,Si weak	−9	***−39***	−30	–	+2
193237	B1 ep	+6	***−97***	−14	+25	+17
209409	B7 Ive	−18	***+53***	+24	–	+4
210839	O6 If(n)p(e)	+6	***−69***	***−55***	–	−19
217891	B6 IIIe	+4	***+65***	***+48***	−1	+4
17769	B7 V	+18	+41	***+49***	–	−3
19374	B1 V β Cep	***+50***	***+62***	***+91***	–	−5
22951	A1 Vn	***+32***	+33	***+33***	–	−7
24554	A1 V	***+32***	−6	−10	–	–
28052	F0 V	***+32***	+17	+19	–	+6
28149	B5 V	***+42***	+28	–	–	
29248	B2 III β Cep	+4	***+44***	+32	–	−9
35671	B5 V SB	+18	***+51***	***+35***	–	−5
35770	B9.5 Vn	***+26***	***+40***	***+33***	–	−4
70011	B9.5 V	−2	***+40***	+22	–	+2
76582	A7 V	***+24***	+18	+25	–	−6
83808	F8-G0III + A7m	+18	+18	***+34***	–	−10
107700	G7III + A3IV	***+29***	+1	+18	–	−4
119765	A0 V	−7	***−50***	***−33***	+1	–
139891	B6V SB2	+11	***+41***	+28	–	–
166182	B2 IV	+6	***+38***	+27	–	−3
178596	F2 IV-V	−3	***−45***	−28	–	−8
188260	B9.5 III	−15	***−47***	***−38***	–	–
188350	A0 III	−13	***−49***	***−42***	–	−8
212120	B6 V, ell. var.	***−45***	−15	***−41***	–	−2
222603	A7 V	+7	***−69***	***−36***	−5	−4

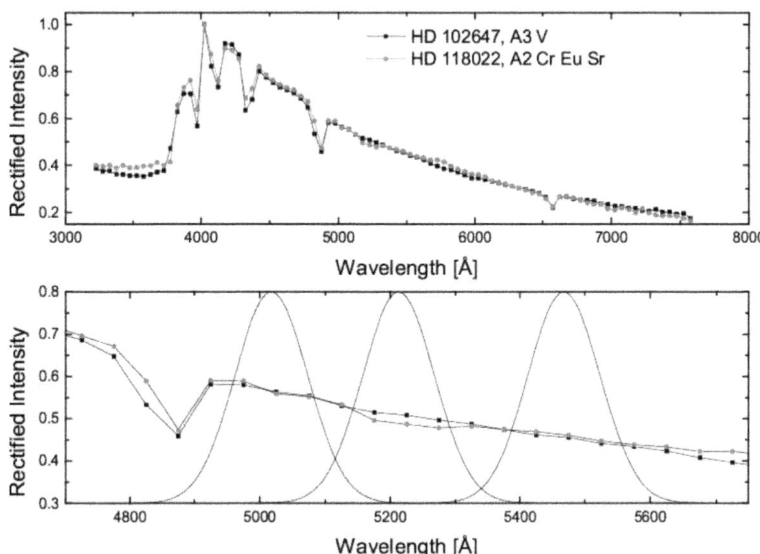

Figure 20: Spectrophotometric data of HD 102647 (A3 V) and HD 118022 (A2 Cr Eu Sr) for the whole spectral range (upper panel) and the region where the Δa system is situated (lower panel). The UV excess and the 5200Å depression, both typical for CP stars, are clearly visible for HD 118022.

- *HD 24554*: Schröder & Schmitt (2007) found strong X-ray emission indicating an undetected binary nature of the object.
- *HD 28052*: This is the second-brightest X-ray source in the Hyades and a possibly quadruple system with at least one component of δ Scuti type (Simon & Ayres, 2000).
- *HD 83808*: This spectroscopic binary system is listed as CP1 candidate in Renson & Manfroid (2009).
- *HD 107700*: Griffin & Griffin (2011) analysed this close-binary system and found a slight metal-weakness for both components.
- *HD 119765*: Renson & Manfroid (2009) listed it as questionable CP candidate without any designation to a specific subgroup.

For these samples, $\Delta a'$ is superior to the other two systems. For almost all of the newly discovered peculiar objects among the normal-type objects, photometric Δa and/or spectroscopic observations are needed to clarify their nature.

In addition, nine cool-type objects ($0.5 < (B-V)_0 < 1.5$ mag) were also identified that significantly deviate in the positive direction. Those stars are

- *HD 19373*: Canto Martins et al. (2011) analysed the chromospheric activity of the G0 V object.
- *HD 26965*: This is a young triple binary system including flare-type objects (Pettersen, 1991).
- *HD 35369*: Prugniel et al. (2011) found an underabundance of [Fe/H] $= -0.22$ dex compared with the Sun.
- *HD 49878*: There are no detailed investigations for this K-type giant available in the literature.
- *HD 68375*: Takeda et al. (2008) investigated it in more detail, and found no peculiarities.
- *HD 82635*: This is a highly chromospherically active RS CVn type giant (Strassmeier et al., 1994).
- *HD 158899*: According to Antipova et al. (2004), this is a moderate Barium star classified as K3.5 III Ba0.1. This group of chemically peculiar stars consists of G- to K-type giants, whose spectra indicate an overabundance of s-process elements.
- *HD 192577*: It is a ζ Aurigae type eclipsing binary of spectral type K4 I (Eaton, 2008).
- *HD 194093*: Gray (2010) presented photometric and spectroscopic time series of this variable star. The variations are found on all time scales up to several hundred days.

The a indices are linearly correlated with $T_{\rm eff}$ up to a spectral type of M0. For cooler-type objects, there is a strong indication of an additional luminosity effect that is superimposed on the temperature dependency.

9.5 Conclusions

The spectrophotometric data of the stellar catalogue by Kharitonov et al. (1988) were used to synthesize three different "a systems". These data cover the complete spectral range from low- to high-mass objects. Excluding low-quality data and early-type supergiants, 1067 stars were used to synthesize different peculiarity indices. The main results are

- Most of the known classical magnetic CP stars were detected.
- A list of about 50 normal-type objects across the complete spectral range was presented that were detected.
- The most efficient a system is very similar to that previously employed by Maitzen (1980b).
- The normality line of the a system correlates with T_{eff} up to a spectral type of M0.

The analysis showed that spectrophotometric data can be used for calculating synthetic a indices and for detecting peculiar objects across the complete spectral range up to M0. These findings are also interesting in the light of the discussion presented in Sect. 8.2 about the 5200Å region of stars across the HRD.

10 A catalogue of Δa measurements for Galactic field stars

The prerequisite for investigating larger samples of CP stars (including the generally fainter star cluster members) is an unambiguous detection. Looking into catalogues of CP stars, especially the magnetic ones, it immediately becomes obvious that there are many discrepancies even at classification dispersions. The reasons for discrepant peculiarity assessments are mainly to be found in the differences in observing material (density of spectrograms, widening of spectra, dispersion, and focussing), but also intrinsic variability of peculiar spectral features (e.g. silicon lines).

Besides the use of (very time consuming and magnitude limited) high dispersion spectroscopy, photometry has shown a way out of this dilemma, especially through the discovery of characteristic broad band absorption features, the most suitable of them located around 5200Å (Kodaira, 1969).

Since its introduction in 1976, many papers presenting Δa photometry of galactic field and especially open clusters have been published. Until the late 90s, all observations were performed with photomultipliers.

All but two (Vogt et al., 1998; Paunzen et al., 2005b) publication about Δa photometry of galactic field stars were explicitly devoted to CP stars

Table 8: The number of individual Δa measurements from the given reference. The used filter systems are according to Table 9.

Ref.	$N_{\rm obs.}$	System	Ref.	$N_{\rm obs.}$	System
Maitzen (1976a)	168	1	Pavlovski & Maitzen (1989)	40	4
Maitzen (1980a)	10	2	Schnell & Maitzen (1994)	3	5
Maitzen (1980b)	8	2	Schnell & Maitzen (1995)	14	5
Maitzen & Segewiss (1980)	21	2	Maitzen et al. (1997)	6	6
Maitzen (1981)	20	2	Maitzen et al. (1998)	131	7
Maitzen & Vogt (1983)	342	3	Vogt et al. (1998)	803	3
Maitzen & Pavlovski (1989a)	16	4	Paunzen et al. (2005b)	99	5
Maitzen & Pavlovski (1989b)	31	4			

or related groups selected on the basis of several relevant catalogues. Superficial "normal" type objects were only observed in order to define the normality line for the specific set of data.

All available Δa data of bright Galactic field stars that have been published in the literature over more than two decades were compiled, in order to make a sound statistical analysis of detection probabilities for all kinds of peculiar objects, as well as luminosity class I/II supergiants (Sect. 11). This is especially important for observations in star clusters (Sects. 17 and 18) and extragalactic systems (Sect. 19). The knowledge of how many peculiar stars for a certain detection limit are expected in comparison with the actually observed number, serves as important information for a statistical analysis, such as the incidence of CP stars that depend on different local environments.

Table 9: Filter systems used for the Δa measurements.

System	g_1		g_2		g_3/y	
	λ_c	$\lambda_{1/2}$	λ_c	$\lambda_{1/2}$	λ_c	$\lambda_{1/2}$
1	5020	130	5240	130	5485	230
2	5010	130	5215	130	5485	230
3	5020	130	5240	130	5505	230
4	5026	111	5238	138	5466	230
5	5017	120	5212	120	5494	228
6	5027	222	5205	107	5509	120
7	5017	110	5212	120	5473	188

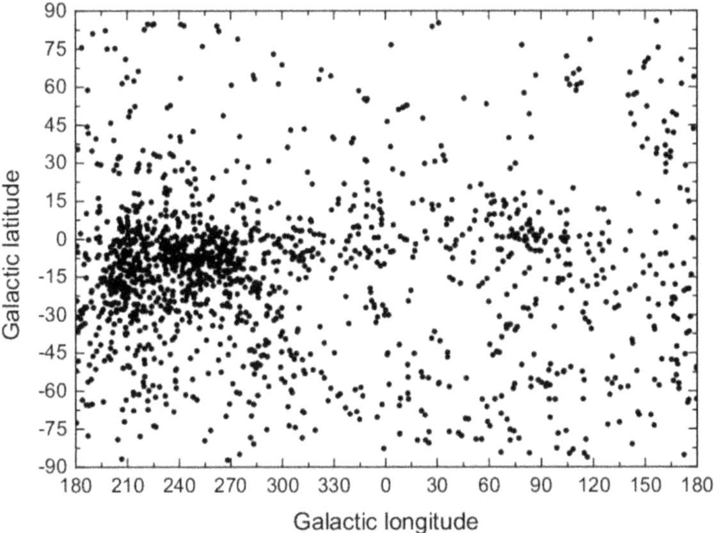

Figure 21: The galactic distribution of the sample including 1561 individual objects. There is a clustering of objects at $190^o < l < 280^o$, the almost reddening free (up to 500 pc from the Sun) third galactic quadrant.

10.1 Selection, preparation, and homogenization of the data

The following sources of Δa photometry for galactic field stars were used: Maitzen (1976a, 1980a,b, 1981), Maitzen & Segewiss (1980), Maitzen & Vogt (1983), Maitzen & Pavlovski (1989a,b), Pavlovski & Maitzen (1989), Schnell & Maitzen (1994, 1995), Maitzen et al. (1997, 1998), Vogt et al. (1998), and Paunzen et al. (2005b). The number of individual measurements are given in Table 8. Some references already list $(b-y)_0$ magnitudes, these were used. Objects were only included which have Strömgren indices available.

All but Maitzen et al. (1997, CCD) were conducted with photomultipliers and classical aperture photometry (one star at one time). The different filter sets of these observations are listed in Table 9. The only important difference is the filter set used by Maitzen et al. (1997) together with the CCD equipment. Although the bandwidths of g_1 and y were exchanged, the overall values of Δa are in very good agreement with those of the photoelectric system.

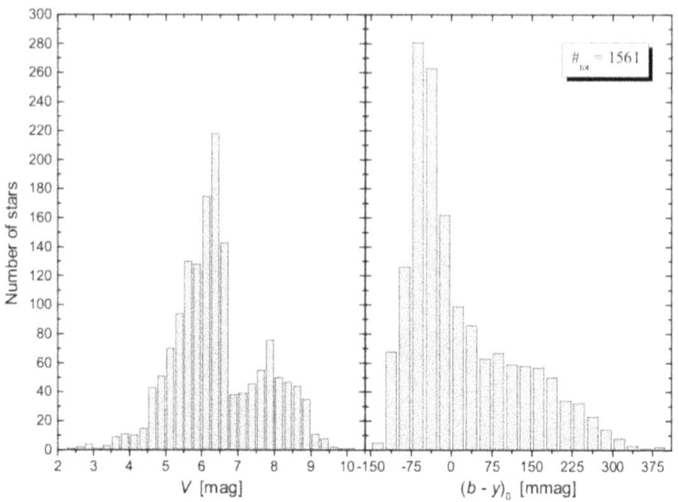

Figure 22: The distribution of Johnson V and Strömgren $(b-y)_0$ of the sample.

The Johnson, Geneva, and Strömgren (if not listed in the original source of the Δa photometry) colours were taken from the General Catalogue of Photometric Data (GCPD, Mermilliod et al., 1997). Usually, the reddening for objects within the solar neighbourhood is estimated using photometric calibrations in the Strömgren $uvby\beta$ system (Strömgren, 1966; Crawford & Mander, 1966; Crawford, 1975, 1979; Hilditch et al., 1983). The validity of these procedures for CP stars was shown, for example, in Netopil et al. (2008). However, these relations have to be used with some caution when applied to magnetic CP2 stars, because all calibrations are primarily based on the β index. But due to variable strong magnetic fields, the β index can give erratic values (Catalano & Leone, 1994) Another effect taken into account is a "blueing" effect, which occurs in bluer colours due to stronger UV absorption than in normal type stars, which can be erratically interpreted as strong reddening (Adelman, 1980). An independent way to derive the interstellar reddening is to use galactic reddening maps, which are derived from open clusters as well as from galactic field stars. For this comparison, the model proposed by Chen et al. (1998) was used to derive the interstellar reddening for all program stars. The values from the calibration of the Strömgren $uvby\beta$ and the model (Chen et al., 1998) agree within an error of the mean of ± 1.9 mmag for the complete sample.

In order to check this procedure, the raw data of Maitzen (1976a) were used to derive the normality line of these data. He used a combined normality line for the $(b-y)$ and $(b-y)_0$ values under the assumption that the reddening in the solar neighbourhood is negligible. The deduced results confirm his A_0 parameter with a slight adjustment of the A_2 value from 0.06 to 0.066, which means a correction of Δa in the order of only one mmag. This leads to confidence in the applied dereddening procedure.

As the next step, the intrinsic consistency of the different reference for objects in common was checked since the filter systems are slightly different (Table 9). Besides a possibly wrong identification or typographical errors, the variability of Δa itself has to be taken into account. Several CP stars show variability in correlation with strong magnetic fields and rotation. These variations were also detected in the Δa photometric system (Sect. 5).

The published $(b-y)_0$ and Δa values were checked in this respect and no significant trends were found. All references agree, in a statistical sense, within one sigma of the correlation coefficients. No significant outliers were found. Therefore, the values of objects with more than one measurement were averaged without using any weights, resulting in 1561 individual galactic field stars. This almost doubles the sample used by Vogt et al. (1998) with an available Δa value. This sample consists of "normal" type and peculiar stars of all kinds.

Figure 21 shows the distribution of the galactic coordinates for the sample. There is a clustering of objects between $190^o < l < 280^o$ (third galactic quadrant) because of the systematic investigation in the southern hemisphere by Vogt et al. (1998). In this region, the reddening is almost negligible for distances up to 500 pc (Chen et al., 1998). But no unknown bias has been introduced, because the CP stars are distributed uniformly in the solar neighbourhood (Gómez et al., 1998).

The sample covers $-0.15 < (b-y)_0 < +0.4$ mag (Fig. 22), which corresponds to the spectral range from B0 to F8 (Crawford, 1975, 1979). The peak of the distribution is at $(b-y)_0 = -0.05$ mag or B9 at the MS. At this spectral type, the number of CP2 stars reaches an intrinsic maximum. The Johnson V magnitude distribution shows two maxima around $V = 6$ mag and 8 mag and extends from $1.5 < V < 10.5$ mag. The first peak is a consequence of the systematic investigation of all bright stars by Vogt et al. (1998) whereas the second one is due to the chosen sample by Maitzen & Vogt (1983). They have investigated the list of astrophysically interesting stars by Bidelman & MacConnell (1973), which was published prior to the Michigan spectral survey, while both are based on the same spectroscopic material.

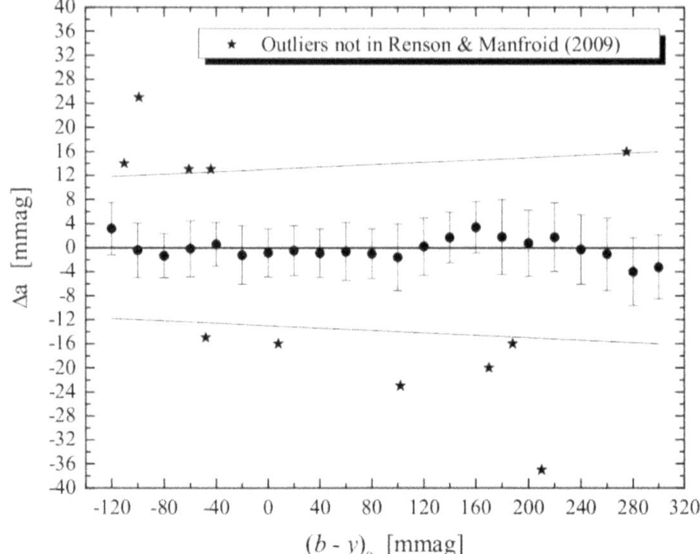

Figure 23: The mean values and the corresponding root mean square scatter for all normal type objects with Δa measurements. The straight lines are the 3σ fits through the errors, whereas the asterisks denote the outliers of objects not included in the catalogue of RM09 discussed in Section 11.2.

11 The detection probability of the Δa system for Population I stars

With the sample of 1561 Galactic field stars (Sect. 10), a statistically sound analysis was performed. There are two stars in the sample that exhibit positive Δa values, for which it was not possible to designate a distinctive membership in a certain class of CP stars. Certainly, these two objects deserve further interest:
HD 68161: Eggen (1980) lists a photometric spectral type of B7III, whereas the spectroscopic type is B8Ib/II (Houk & Swift, 1999). This is a strong indication that this star is of CP2 type, because the silicon lines are also used to derive the luminosity classification.
HD 225253: several deviating classifications in the literature were found, B8III (Jaschek et al., 1969), B8IV/V p(mild) (Maitzen, 1980a) and B7 UV gallium (Jaschek & Jaschek, 1987). With a Δa value of +18 mmag, this object is obviously chemically peculiar.

Let us recall the definition of the different (quality) flags for CP stars by Renson & Manfroid (2009, RM09 hereafter). In general, a flag indicates the degree of confidence for the CP nature of the star. A slash (/) denotes a star that was improperly considered to have an Ap, HgMn or Am nature. Question marks (?) mean doubtful cases (2314 stars, among which are 311 Ap stars detected by photometric criteria, and not yet confirmed by ordinary spectroscopy). On the contrary asterisks (*) denote well-known confirmed Ap, HgMn or Am stars (only 426).

For the further analysis of the CP stars, only well-known confirmed objects were used. This is the most homogeneous sample of CP stars available in the literature. Of course, one could add further stars to this list on the basis of more recent published single papers. But this would decrease the homogeneity of the sample. However, a list of further confirmed CP stars on the basis of the Δa index was derived (Table 11). These stars are already included in RM09, but without an asterisk flag. They should be included with such a flag in the next updated version of RM09.

11.1 The normal type objects

The selection of apparent normal type objects is a rather difficult task because of undetected peculiarities of all kinds which might introduce an unknown bias. All objects not listed in the catalogue by RM09 were chosen, because they compiled stars which have been identified as peculiar at least once in the literature, even though they might turn out to be "normal" after all. As a next step, stars classified as λ Bootis (Paunzen et al., 2002a), Be/shell stars, and super giants of luminosity classes I and II were excluded. For this purpose the spectral classifications given in the Michigan catalogues of two-dimensional spectral types (Houk & Swift, 1999, and references therein) and the extensive list of Skiff (2016) were used. Known binary systems of all kinds were not automatically excluded.

In total, 673 objects with luminosity classes V, IV, and III were selected on this basis. These stars were divided into subsamples according to their $(b-y)_0$ values ($-0.125 < (b-y)_0 < +0.315$ mag) with a bin size of 20 mmag, which corresponds to ± 1 spectral subclasses. The intervals are defined in the sense [a,b]. For each subsample the mean Δa value and the corresponding standard deviation of the mean for all objects within it was calculated. One object, HD 102870 (F8 V), falls outside the above given bins because it is too cool with $(b-y)_0 = +0.356$ mag. Figure 23 shows the results, with mean values uniformly distributed around zero. The straight lines $\pm 13 \pm 0.01(b-y)_0$ are linear fits through the mean values adding 3σ in both directions. The mean value of σ is 5(2) mmag in the interval from $-0.125 < (b-y)_0 < +0.315$ mag which corresponds to spectral types

Table 10: Limits in mmag for detecting the individual groups of objects. For example 41% of all CP1 stars can be detected with a 3σ limit of +5 mmag. In addition, the mean group $\overline{\Delta a}$, the maximum values for group members Δa_{m}, and the number N of well-established members for the analysis are listed.

	+5 [%]	+10 [%]	+15 [%]	+20 [%]	$\overline{\Delta a}$ [mmag]	Δa_{m} [mmag]	N
CP1	41	21	3	3	+3.5	+22	29
CP2	96	92	85	71	+31.3	+79	129
CP3	56	19	–	–	+4.3	+11	16
CP4	79	71	43	21	+18.8	+31	14
I/II	52	17	7	–	+4.0	+19	29
	−5 [%]	−10 [%]	−15 [%]	−20 [%]			
λ Boo	95	65	35	20	−16.2	−35	20
Be	20	5	2	–	−0.5	−19	64

between B0 and F5. This is in line with the published values in Maitzen (1976a) and Vogt et al. (1998). The detection limit for the sample increases from ±12 to ±16 mmag towards cooler objects (Fig. 23 and Sect. 9). There are two main reasons for this behaviour: 1) $(g_1 - y)$ is no longer only an indicator of the T_{eff} because of significantly increasing line blanketing as $(b - y)$ and 2) the strong general increase in metallic lines towards later spectral types. Since most of the "normal" type stars are brighter than $V = 7$ mag (Fig. 22), it was checked if the detection limit changes if only fainter "normal" objects are used. Although one is confronted with poor number statistics, no correlation was found.

The observed scatter around the normality line has several causes. Besides the instrumental limitations (e.g. photon noise and variable sky transparency), which account for about 3 mmag (Maitzen, 1976a), the intrinsic variations due to the natural bandwidth of the main sequence is important. In Sect. 14, a synthetic Δa photometric system based on modern stellar atmospheres and the filter transmission curves of "System 3" in Table 9 was investigated. The models are in the temperature range from 7000 to 15000 K and surface gravities from 2.5 to 4.5 dex. A natural bandwidth of about 4 mmag within these models was found. A metallicity range from −0.5 to +0.5 dex creates an overall bandwidth (= 2σ) of 10 mmag (Sect. 14.1.4). This is more or less exactly the value one finds for the sample of normal type, galactic field stars. In open clusters for which the metallicity is rather uniform, lower detection limits are expected if neglecting strong differential reddening (Sect. 17).

The results of the statistics for the different groups are listed in Table 10. The detection rates as functions of the corresponding 3σ limits are shown in Figure 24.

11.2 Outliers not included in RM09

Eight objects (four each with positive and negative Δa values) are located outside the 3σ limit (Fig. 23) and are not included in RM09.

First of all, the four objects with negative Δa values are discussed in more detail:
HD 2834 ($(b-y)_0 = +8$ mmag, $\Delta a = -16$ mmag): a known spectroscopic binary system that is also part of a visible double system.
HD 4158 (+210, −37): a metal weak, cool star classified as hF3mF0V (wk met) by Paunzen et al. (2001).
HD 6173 (+102, −23): although previously thought to be a member of the λ Bootis group, it was classified as A0IIIn (Paunzen et al., 2001).
HD 97937 (+188, −16): a metal weak (F0V m−1.5), intermediate Population II type object (Gray, 1989).
HD 114911 (−48, −15): a young, early type quadruple system which exhibits X-ray emission (Hubrig et al., 2001).

The four objects with positive Δa values lie only marginally above the 3σ limit (Fig. 23):
HD 31726 (−110, +14): a B1V star which exhibits rather normal silicon line strengths (Massa, 1989).
HD 49028 (−61, +13): a close visual binary system with a B7III and a F3V Fe−2 component (Corbally, 1984).
HD 65908 (−44, +13): Bowyer et al. (1995) reported a clear flux detection in the far-ultraviolet for this star which might indicate a previously unknown binary nature.
HD 73451 (+275, +16): a known spectroscopic binary system with an early A- and a G-type component (Cowley et al., 1969).

These objects deserve a further analysis using photometric and spectroscopic data to determine why they exhibit a non-standard Δa value.

11.3 CP1 stars

In the investigated sample, there are only 29 well established CP1 stars yielding a slightly positive mean value (+3.5 mmag) and quite extreme

Table 11: Confirmed CP stars listed in RM09 with the $(b-y)_0$ and Δa values taken from Section 10.

HD	RM09	Spec.	$(b-y)_0$ [mmag]	Δa [mmag]	HD	RM09	Spec.	$(b-y)_0$ [mmag]	Δa [mmag]
315	30	B9 Si	−71	+31	47116	12650	B9 Si	−50	+35
2957	760	B9 Cr Eu	−20	+36	47714	12810	B8 Si	−64	+35
5601	1480	A0 Si	−56	+49	47802	12860	B9 Si	−60	+30
6164	1580	A0 Si	−12	+40	48729	13000	A0 Si	−35	+26
6322	1620	A0 Sr Cr Eu	−23	+22	50166	13690	A0 Si	−22	+30
6783	1760	B8 Si	−57	+53	50221	13750	B9 Si	−69	+28
8783	2110	A2 Sr Eu Cr	+72	+23	50403	13790	A2 Sr Eu	+98	+20
16145	4060	A0 Cr Sr Eu	+28	+35	50461	13810	B9 Si Cr	−51	+52
18610	4670	A2 Cr Eu Sr	+114	+48	50825	13960	B8 Si	−29	+36
19712	4880	A0 Cr Eu	−60	+43	51172	14100	B9 Si	−58	+33
21590	5400	B9 Si	−43	+31	51650	14210	B9 Si	−4	+33
23207	5900	A2 Sr Eu	+106	+28	51684	14220	F0 Sr Eu Cr	+154	+20
27463	7050	A0 Eu Cr	+22	+28	52589	14400	B8 Si	−62	+42
28299	7230	B8 Si	−57	+31	52696	14460	A3 Sr Eu Cr	+97	+23
28365	7280	B9 Si	−71	+18	52847	14500	A0 Cr Eu	+102	+62
29435	7530	B9 Si	−46	+22	52993	14550	B9 Si	−64	+29
29925	7690	B9 Si	−61	+46	53116	14620	A0 Sr Eu	−46	+50
32145	8190	B7 Si	−72	+37	53204	14630	B9 Si	−30	+21
32432	8250	B9 Si	−32	+31	53662	14740	A0 Si	−36	+19
34427	8760	A0 Si	−14	+25	53851	14780	B9 Si	−63	+29
34435	8770	A2−A7	+70	+17	54832	14980	B9 Si	−64	+42
34631	8840	B9 Si	−64	+24	55309	15040	B9 Si	−60	+48
34719	8850	A0 Hg Si Cr	−38	+44	55395	15070	B9 Si	−65	+22
34736	8860	B9 Si	−64	+17	55540	15100	A0 Eu Cr	−68	+70
35353	9030	A0 Sr Cr Eu	+94	+23	56022	15250	A0 Si Cr Sr	−25	+16
36668	9560	B7 He wk. Si	−59	+26	56273	15300	B8 Si	−59	+25
37140	9910	B8 Si Sr	−66	+31	56336	15350	B9 Si	−63	+32
37189	9940	A0 Si	−9	+23	56632	15450	B9 Si	−41	+22
37210	9950	B8 He wk. Si	−61	+22	56809	15480	B9 Sr Cr Eu	−3	+29
37642	10150	B9 He wk. Si	−84	+34	56882	15520	F0 Sr Cr Eu	+109	+27
37713	10170	B9 Si	−64	+15	57526	15710	B9 Si	−38	+40
37808	10210	B9 Si	−73	+27	58292	15910	A0 Si	−27	+34
38170	10256	B9	−33	+23	60559	16550	B8 Si	−62	+23
38698	10400	A0 Si	−13	+54	61382	16840	B9 Si	−23	+23
38719	10410	A0 Cr Sr Eu	+11	+38	62530	17150	B9 Eu Cr	−31	+34
39082	10500	B9 Sr Cr Eu	−63	+42	62535	17160	A0 Si	−35	+40
39353	10570	B8 Si	−57	+17	62556	17180	A0 Eu Cr	+35	+18
39575	10600	A0 Si Cr Eu	−74	+55	62640	17220	B9 Si	−80	+19
40071	10680	B9 Si	−50	+42	63401	17480	B9 Si	−73	+27
40277	10730	A1 Sr Cr Eu	+41	+20	63759	17570	A2 Sr Cr Eu	+51	+22
40383	10780	B9 Si	−42	+31	64881	17800	A0 Si	−55	+32
40711	10880	A0 Sr Cr Eu	+95	+43	64901	17810	A0 Si	−63	+38
40759	10900	A0 Cr Eu	−13	+27	66051	18190	A0 Si	−29	+30
41403	11080	B9 Sr Cr Eu	−16	+48	66195	18240	A0 Sr Eu Cr	+43	+39
42326	11280	A0 Eu Cr	+8	+42	66350	18320	A0 Eu Cr	−30	+37
42335	11290	A0 Si	+72	+16	66624	18410	B9 Si	−85	+28
42536	11340	A0 Sr Cr	−4	+22	66698	18430	A0 Eu	−26	+21
42576	11360	B9 Si Cr	−16	+20	67165	18510	A0 Si	−44	+36
43408	11520	B9 Sr Eu	−23	+28	67330	18590	A0 Si	−13	+38
43901	11660	F0 Sr Cr Eu	+132	+26	67835	18700	B8 Si	−65	+17
44290	11750	B9 Cr Eu	−26	+36	68161	18840	B8	−44	+22
44293	11760	A0 Cr	−89	+55	68292	18860	B9 Si	−73	+35
44456	11800	A0 Si Sr	−9	+50	68476	18960	B9 Si	−53	+22
44947	11910	A3 Sr Eu Si	+105	+21	68998	19120	A5 Eu Cr Sr	+130	+18
44953	11930	B8 He wk.	−86	+17	69067	19150	B8 Si	−62	+39
45439	12040	B9 Si	−60	+19	70464	19440	B9 Si	−68	+30
45530	12070	A0 Si	−30	+36	70507	19470	B9 Si Cr	−60	+24
45583	12120	B9 Si	−65	+58	70749	19520	B8 Si	−54	+23
45698	12150	A2 Sr Eu	+69	+16	70847	19570	B8 Si	−66	+16
46462	12430	B9 Si	−75	+25	71808	19820	B9 Si	−54	+36

Table 11: continued.

HD	RM09	Spec.	$(b-y)_0$ [mmag]	Δa [mmag]	HD	RM09	Spec.	$(b-y)_0$ [mmag]	Δa [mmag]
72055	19900	B8 Si	−58	+23	98340	28350	B9 Si	−38	+19
72295	19980	A0 Sr Eu Cr	−23	+35	98457	28360	A0 Si	−39	+22
72611	20120	A0 Eu Cr Sr	−62	+49	98486	28370	B9 Si	−60	+23
72634	20130	A0 Eu Cr Sr	−11	+26	101600	29270	A0 Si	+22	+21
72881	20200	B9 Si	−61	+35	101724	29330	B9 Si	−78	+16
72976	20250	B9 Si	−58	+37	103302	29780	A0 Sr Cr Eu	+4	+33
73737	20560	B9 Si	−25	+46	103457	29820	A0 Si	−7	+35
74067	20630	A0 Cr Si	−50	+37	103498	29830	A1 Cr Eu Sr	−9	+46
74555	20830	A0 Cr Eu	+6	+23	104810	30330	B8 Si	−65	+19
74888	20950	B9 Si	−64	+22	105379	30460	A0 Sr Cr	+27	+16
76104	21290	B9 Si	−39	+39	105457	30490	A0 Si	+56	+50
76439	21510	B9 Si	−69	+19	105770	30610	B9 Si	−68	+18
76650	21610	B9 Si	−64	+25	106204	30700	B9 Si	−46	+25
76897	21690	B9 Si	−32	+25	109809	31860	B9 Si	−48	+24
77609	21960	B9 Eu Sr	−29	+32	112252	32600	B9 Si	−42	+19
77653	21970	B9 Si	−70	+26	112528	32730	A3 Sr Eu Cr	+188	+24
77689	21980	B9 Si	−77	+38	115440	33340	B9 Si	−61	+42
78201	22150	A0 Sr Eu	−27	+23	115599	33420	A2 Si	+158	+22
78568	22260	B9 Si	−62	+16	116235	33550	A2−A8	+52	+16
79066	22420	A9− dD	+202	−16	116423	33580	A0 Eu Sr Si	+87	+48
79976	22750	B9 Sr Cr Eu	−15	+16	116890	33720	B9 Si	−68	+17
80282	22830	A0 Si	−73	+30	117057	33780	B9 Si	−69	+17
81141	23030	B9 Si	−58	+25	118913	34340	A0 Eu Cr Sr	+17	+44
81289	23080	A2 Eu Sr Cr	−5	+25	119308	34460	B9 Sr Cr Eu	−26	+36
81588	23190	A5 Sr Cr Eu	+118	+18	120059	34630	B8 Si	−15	+15
81847	23260	B8 Si	−65	+47	121661	34970	A0 Eu Cr Si	+27	+39
82093	23340	A2 Sr Eu Cr	+5	+41	122208	35050	A2 Sr Cr Eu	+45	+42
82154	23360	B9 Si	−51	+59	123627	35410	A3 Sr Eu Cr	+144	+20
82567	23480	B9 Si	−62	+26	125532	35850	B8 Si	−67	+38
83266	23750	A0 Si Cr Sr	−48	+55	126876	36120	B8 Si	−62	+20
83625	23850	A0 Si Sr	−67	+40	127453	36240	B8 Si	−57	+33
85892	24520	B8 Si	−61	+27	127575	36280	B9 Si	−60	+48
86170	24620	A2 Sr Eu Cr	+71	+23	129899	37030	A0 Si	−26	+44
86976	24850	A5 Sr Eu Cr	+133	+20	130335	37110	A2 Si	+120	+33
88385	25310	A0 Cr Eu Si	−2	+50	131505	37350	B9 Si	−68	+18
89103	25530	B9 Si	−54	+59	132322	37580	A7 Sr Cr Eu	+118	+25
89192	25580	A0 Cr Eu Si	+9	+27	133281	37840	B9 Si	−63	+38
89217	25590	B9 Si	−50	+26	135415	38460	B8 Si	−61	+31
89385	25650	B9 Cr Eu Si	−26	+40	137160	38970	A0 Sr Eu Cr	+47	+25
89393	25660	A0 Sr Cr Eu	+134	+42	137193	38980	B9 Si	−46	+39
89519	25720	B9 Eu Cr Sr	−33	+24	138758	39500	B9 Si	−38	+53
89680	25750	A0 Cr	+33	+21	138773	39520	A0 Si	+81	+27
90264	25960	B8 He wk.	−68	+17	141461	40170	B9 Si	−60	+29
90612	26020	B8 Si	−60	+47	141641	40220	B8 Si	−78	+21
91089	26200	B9 Si	−69	+16	143473	40620	B9 Si	−67	+45
91134	26210	B9 Si	−48	+29	144231	40860	B9 Si	−22	+42
91239	26240	A0 Eu Cr Si	−33	+44	144748	41020	F0 Sr Eu Cr	+85	+35
92379	26610	B8 Si	−56	+35	146971	41510	A0 Sr Cr Eu	+147	+16
93500	27010	B9 Cr Eu Sr	−19	+15	148848	42060	A0 Si Cr Sr	+134	+51
93821	27150	B9 Si	−40	+16	149764	42360	A0 Si	+10	+21
94455	27260	A0 Cr Eu Sr	+112	+38	149831	42400	B9 Si	−70	+22
94873	27350	B8 Si	−49	+19	150040	42450	A0 Si	+37	+20
95198	27420	B9 Si	−34	+31	150323	42500	B6 Si	−61	+40
95413	27490	A0 Si	−52	+44	150486	42560	B9 Si	−40	+22
95442	27500	A0 Sr Cr Eu	+13	+24	150714	42640	A0 Si	+88	+24
95569	27530	B9 Si	+121	+29	151742	42930	B9 Si	−51	+25
95699	27560	F0 Sr Eu Cr	+75	+36	151965	43000	B9 Si	−78	+35
96910	27960	B9 Si Cr Eu	−24	+61	152366	43130	B8 Si	−53	+42
97394	28090	A5 Eu Cr Sr	+121	+39	153707	43400	B8 Si	−37	+20
97986	28260	B8 Si	−48	+30	154253	43610	A0 Sr Cr Eu	+106	+20

Table 11: continued.

HD	RM09	Spec.	$(b-y)_0$ [mmag]	Δa [mmag]	HD	RM09	Spec.	$(b-y)_0$ [mmag]	Δa [mmag]
154308	43620	A0 Cr Eu Sr	+95	+47	174779	48940	A0 Si	−45	+49
154458	43660	B9 Si	−66	+35	176555	49260	B9 Si	−58	+24
155127	43810	B9 Eu Cr Sr	−14	+45	181550	50290	B9 Si Cr	−39	+31
155778	43990	B8 Si	−40	+24	184020	50800	A0 Sr Cr Eu	−15	+18
156300	44040	B9 Si	−34	+29	189502	52540	A0 Si Sr Cr	−48	+39
156853	44120	B8 Si	−58	+31	191439	53340	B9 Cr Eu Sr	−25	+46
156869	44130	A0 Sr Cr Eu	+41	+25	191796	53460	A0 Eu Cr	−35	+39
157678	44280	B9 Si	−66	+16	196425	54740	A7−F2	+121	+17
157751	44320	B9 Si Cr	−29	+44	196606	54840	B9 Si	−55	+20
158128	44450	B9 Si	−41	+36	196655	54880	A2−	+46	+34
158175	44480	B8 Si	−55	+42	197417	55030	A0 Cr Eu	+32	+54
158450	44620	A0 Sr Cr Eu	+234	+22	200405	55830	A2 Sr Cr	+13	+38
159545	44900	B9 Si	−47	+31	200623	55940	A2 Sr Eu Cr	+67	+25
159846	44970	B9 Si	−58	+23	203585	56690	A0 Si	−37	+21
160127	45090	A2 Sr Eu Cr	+128	+31	204131	56860	B9 Si Cr Sr	−10	+15
161277	45440	B9 Si	−42	+33	204541	56970	A2−F1	+138	+17
161349	45470	B9 Si	−67	+16	204815	57030	A0 Si	+16	+40
164224	46430	B9 Cr Eu	−31	+44	207188	57640	A0 Si	−41	+37
166053	46770	B9 Si	−73	+16	209339	58240	B0 He	−120	−26
166921	46960	B8 Si	−47	+28	209845	58400	A1−F1	+79	+15
168856	47330	B9 Si	−63	+38	212432	58920	B9 Si	−54	+22
169021	47410	B9 Si	−63	+27	213232	59100	A4 Sr	+63	+17
172690	48380	A0 Si Sr Cr	+24	+18	215966	59650	B9 Eu Cr	−13	+31
173406	48570	B9 Si	−59	+24	217792	60045	F0−	+191	−15
173562	48600	A0 Cr Eu	+38	+21	225253	61770	B8 Ga	−46	+18
174638	48890	B8 He	−75	+36	258583	12260	A0 Si	−20	+18
174646	48910	B9 Si	−58	+25		17100	A0 Si	−19	+46

values (+22 mmag) for some members (e.g. HD 184552). But the detection capability is only 21% for a limit of +10 mmag (Table 10). Some of the most outstanding objects have already been discussed by Vogt et al. (1998). In Sect. 14, detailed synthetic Δa values for the CP1 group are presented concluding that the 5200 Å flux depression is only marginally detectable for these objects. This means that the most significant elements contributing to this feature (e.g. Iron and Chromium) are not strongly enhanced compared to the overall strength of the metallic-lines in CP1 stars.

11.4 CP2 stars

Most of the former Δa observations were dedicated to this group because of the high efficiency at detecting CP2 stars. This is also reflected by the results listed in Table 10. About 92% of all CP2 objects can be detected with a limit of +10 mmag, whereas the mean value is +31.3 mmag with an extreme value of +79 mmag. This sample includes all well-established CP2 stars classified in RM09 and marked with an asterisk. The sample was not subdivided into hotter Si and cooler CrEu(Sr) objects, because the definition is not quite clear yet (Bychkov et al., 2003).

11.5 CP3 stars

All CP3 stars listed by Adelman et al. (2003) with Δa measurements (10 objects) are included in the analysis. The mean value of all objects is +4.3 mmag but with a very low maximum of +11 mmag (Table 10). The detection limit drops from 56% to 19% for +5 and +10 mmag, respectively. This is comparable to the values found for the CP1 group. This similar behaviour can be explained because both groups are apparently non-magnetic.

11.6 CP4 stars

The percentage of detection (71%) at +10 mmag is lower than that of the CP2 group. But one can conclude that a statistically significant amount of magnetic CP stars can be detected with the Δa photometric system.

Within the sample, there are also three helium rich objects: HD 37017 ($\Delta a = +2$ mmag), HD 37479 (+10), and HD 64740 (−1). Zboril et al. (1997) analysed a sample of 17 helium rich objects and concluded that several stars exhibit strong emission together with stellar activity. This might be the reason why such a wide range of Δa values was observed similar to Be/shell stars.

11.7 λ Bootis stars

The group of λ Bootis stars is an especially excellent example of how Δa photometry can preselect candidates for spectroscopic observations, for example, in young open clusters. Paunzen et al. (2001) presents spectral classification of 708 stars selected to be good photometric candidates only on the basis of Strömgren indices. From those, only 26 turned out to be new members of the λ Bootis group.

Within the presented analysis, twenty well-established members of the λ Bootis group with Δa measurements were found. The group mean value is −16.2 mmag, and a maximum of −35 mmag (Table 10) shows the high efficiency of this photometric system. Even with a detection limit of −10 mmag, almost 2/3 of all bona-fide λ Bootis stars can be detected.

11.8 Supergiants

The only notice about a positive detection of supergiants was given by Vogt et al. (1998), who investigated two cool supergiants and found a substantial positive deviation from the normality line.

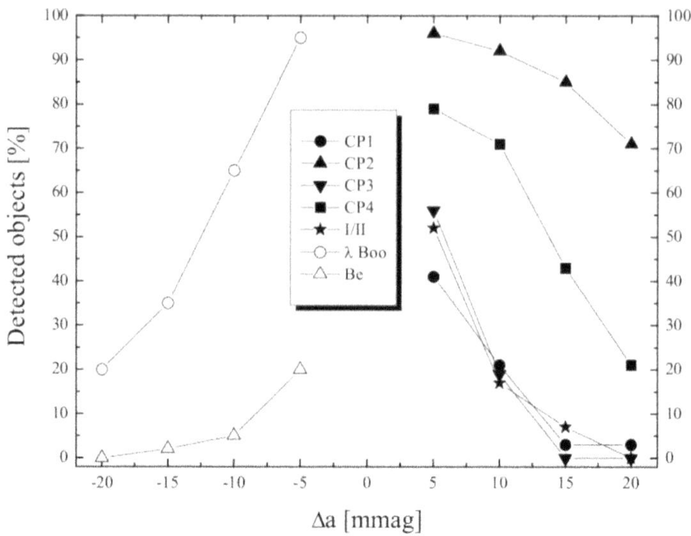

Figure 24: The detection probability of the different investigated groups as listed in Table 10.

The current analysis was restricted to objects classified as luminosity class I or II in the literature. It has to be emphasized that such objects are, in general, easily sorted out within colour-magnitude diagrams of different photometric systems. However, the results from Sect. 13 show that isochrones with the Δa photometric system together with the location of objects with respect to the normality line are capable of sorting out fore- and background objects very efficiently.

The analysis of supergiants in open clusters is important in several respects. Most of these objects are within binary systems and exhibit variability on several time scales (Corliss et al., 2015). Furthermore, their membership is crucial for isochrone fitting because of the sensitivity of the determined age on the existence of a "giant clump" (Eigenbrod et al., 2004). Since all known CP stars have luminosity classes IV or V (Gómez et al., 1998), supergiants selected by their location in a CMD with a significant positive Δa value can be easily tested for membership in an open cluster (Sect. 17.3). In total, 29 supergiants from O9.5II (HD 47432) to F4Iab (HD 61715) are included in the investigated sample with a mean value of +4.0 and a maximum value of +19 mmag (Table 10).

11.9 Be/shell stars

The phases of emission are replaced by shell and normal phases of the same object. This episode was analysed for the case of Pleione using Δa photometry (Sect. 5.2). In the shell phase it reached a Δa value of +36 mmag, which dropped to +4 mmag within one year. However, the behaviour of Pleione seems quite extreme and outstanding, because no other similar object has been detected so far (Vogt et al., 1998). The contamination of classical CP stars due to Be stars in a shell phase is, therefore, only marginal (Sect. 5.2).

Since Pavlovski & Maitzen (1989) already presented a paper with measurements of 40 apparent Be/shell type stars, the investigated sample is rather large, 64 objects in total. The mean value is close to zero (−0.5 mmag) with extremes of −19 mmag (emission phase) and +36 mmag (shell phase).

11.10 Newly confirmed CP stars

In Sect. 11.1, the detection limit of apparent peculiar was estimated as increasing from ±12 to ±16 mmag towards the coolest objects (spectral type F5). These values are only valid for the sample of 1561 Galactic field stars as presented in Section 10.

Taking a limit of ±15 mmag for the whole investigated range, 294 peculiar objects were detected. This stars together with the spectral information from RM09 are listed in Table 11. There are only three stars with significant negative Δa values among the sample:

- HD 79066: This star was classified as δ Del type (Cowley, 1973) which are a group of giants of late A- to early F-type, with weak Ca II lines.
- HD 209339: classified as O9.7 IV by Sota et al. (2014) which probable shows emission lines.
- HD 217792: a metal-weak binary system (Gray & Garrison, 1989) with a debris disk (Chen et al., 2014).

Only six bona-fide CP1 objects (the references of the classification are listed below) with a significant positive Δa values were found:

- HD 34435: Jaschek & Jaschek (1960)
- HD 116235: LeBlanc et al. (2015)

- HD 196425: Abt et al. (1979)
- HD 196655: Bidelman (1985)
- HD 204541: Bertaud (1970)
- HD 209845: Gray et al. (2003)

These stars are good candidates to search for an organized and stable stellar magnetic field in CP1 objects.

Besides HD 68161 and HD 225253 which are discussed in Sect. 11, six He-peculiar objects (HD 36668, HD 37210, HD 37642, HD 44953, HD 90264, and HD 174638), and only one marginal CP3 star (HD 34719) are listed in Table 11. The latter is a very peculiar object with a strong magnetic field (Kudryavtsev et al., 2006).

As expected, the large majority of objects (275) with significant positive Δa values are CP2 stars. They are included in RM09 but not marked as "well established". These objects are obviously magnetic CP stars (Sect. 14.9.2). Although several of these stars have already been assigned to the CP2 group, the results can be seen as further proof of membership.

11.11 Comparison with the Geneva $\Delta(V1-G)$ and Z indices

These two indices ($\Delta(V1-G)$ and Z) are described in Section 4. Let us recall that Z is only useful defined for stars hotter than A0.

The samples for all star groups including the normal type objects as described before (Table 10) were taken and searched for all objects with available Geneva photometry in the General Catalogue of Photometric Data (GCPD, Mermilliod et al., 1997). In the following, only stars were included in the analysis for which both, Δa and Geneva photometry, is available. The detection limit for the Δa photometric system was set to ±15 mmag which is a very conservative value based on the statistics listed in Sect. 11.1.

Hauck & North (1982) concluded that apparent normal stars have mean $\Delta(V1-G)$ values of -5 mmag and are not uniformly distributed around zero. Therefore, first it was checked if this offset is also visible for the sample as defined in Section 11.1. In total, 635 stars are included in this sample. As for the Δa photometry, the $\Delta(V1-G)$ values were binned according to their $(b-y)_0$ values ($-0.125 < (b-y)_0 < +0.315$ mag) with a bin size of 20 mmag. Only very few normal type stars exceed $\Delta(V1-G) > +2$ mmag, whereas many normal type objects have values lower than -10 mmag with

Table 12: A comparison of the detection capability of the Δa and Geneva $\Delta(V1-G)$, as well as Z indices for the objects with available Geneva 7-colour photometry. No objects of the CP1 and λ Bootis group are within the range of valid Z values (defined for stars hotter than A0). One has to keep in mind that the sign of Z is opposite which means that CP stars have significant negative Z values.

	N_{tot}	$N_{\Delta(V1-G)}$		$N_{\Delta a}$		$N_{\Delta(V1-G)}$	$N_{\Delta a}$
		−	+	−	+	[%]	[%]
CP1	28	–	1	–	1	4	4
CP2	129	2	101	–	110	80	85
CP3	16	–	–	–	–	–	–
CP4	13	–	4	–	5	31	38
I/II	29	2	–	–	2	7	7
λ Boo	17	11	–	12	–	65	71
Be	59	1	2	1	1	5	3
normal	635	7	3	4	1	2	1

	N_{tot}	N_Z		$N_{\Delta a}$		N_Z	$N_{\Delta a}$
		−	+	−	+	[%]	[%]
CP2	78	71	–	–	71	91	91
CP3	15	–	–	–	–	–	–
CP4	13	3	–	–	5	23	38
Be	59	5	4	1	1	16	3
normal	354	5	4	1	–	3	1

a mean value of −7.3 mmag for the complete sample. The mean value of σ in the complete interval is 4(1) mmag. Therefore, significance limits of +5 and −19 mmag for $\Delta(V1-G)$ and the complete range were taken.

For Z, only the range $-0.125 < (b-y)_0 < -0.005$ mag (approximately stars hotter than A0) was investigated in the same way. The mean value was determined as +1.3 mmag with a mean value of σ of 4(1) mmag which yields a detection limit of ±12 mmag. One has to keep in mind that the sign of Z is opposite which means that CP stars have significant negative Z values. Table 12 shows the results according to the defined detection limits and the samples.

The incidence of apparent normal type stars which are still outside the detection limits is comparable low (1 to 3%).

The Δa and $\Delta(V1-G)$ indices are able to detect all investigated star groups with more or less the same statistical significance. The slope for the complete sample of CP2 stars in a Δa versus $\Delta(V1-G)$ diagram is $+0.73(5)$ and a zero point of $-7(2)$ mmag as well as $-0.88(6)$ for Z, respectively.

For λ Bootis stars the detection capability is similar to that of Δa. But for the Be/shell stars, the $\Delta(V1-G)$ and Z indices are even more sensitive.

12 An empirical temperature calibration for the Δa photometric system

One of the most important observational diagnostic tools of astrophysics is the HRD allowing to study the correlation between the effective temperature and the absolute magnitude (or luminosity) of astronomical objects. Virtually all stellar astrophysical models are tested according to the HRD.

The absolute magnitudes of stellar objects can be derived directly via parallax measurements, appropriate photometric indices (e.g. Strömgren β) or on a statistical basis in open and globular clusters. The errors of such combined estimates are already rather small (<0.1 mag).

As a consequence, an empirical effective temperature calibration of main sequence, luminosity class V to III B-, A- and mid F-type stars for the Δa photometric system is established. It provides the index (g_1-y) which shows an excellent correlation with $(B-V)$ as well as $(b-y)$ and can be used as an indicator for the effective temperature. This is supplemented by a very accurate colour-magnitude diagram, y or V versus (g_1-y), which can be used, for example, to determine the reddening, distance and age of an open cluster.

A division between the calibration for B-type and cooler stars was done because of the widely different available reddening free and temperature sensitive indices in the the Strömgren $uvby\beta$, Geneva 7-colour, and Johnson UBV systems. In addition, for B-type stars: 1) there are reddening free parameters available allowing to check the dereddening procedure for (g_1-y) and 2) the stellar interiors for these objects are very similar over the whole spectral range. For A-type to early F-type objects, on the other hand, calibrations have to be more sophisticated due to the increase of line blanketing and luminosity effects.

The photometric calibration of effective temperatures is still a very tricky business with several pit falls. The applied calibrations do severely depend on the investigated spectral range and the physics introduced in the models. Smalley & Kupka (1997) give an excellent overview of how convection, for example, can lead to significant deviations for A-type and cooler stars. The statistical calibration of effective temperatures via photometric indices has been done since the introduction of photometric systems (Johnson, 1958; Strömgren, 1966, and references therein). The applied calibrations become more precise and sophisticated as the theoretical stellar atmospheres and the input physics became more realistic. Furthermore, the amount of available photometric and spectroscopic data increases constantly.

With the established intrinsically consistent, empirical, effective temperature calibrations for B-type to mid F-type stars, it is now possible to study individual objects in very distant galactic open clusters and extragalactic systems.

12.1 The B-type stars

In Sect. 14, a synthetic photometric Δa system is established which confirmed the observed dependency of the a index as a function of various colour indices sensitive to the effective temperature and surface gravity variations within the Strömgren $uvby\beta$ and Johnson UBV photometric systems using fluxes from ATLAS9 model atmospheres as well as most recent atomic line data together with opacity distribution function for individual chemical compositions. Furthermore, in Sect. 13 isochrones are presented, taking into account mass loss during the main sequence evolution on the basis of modern equations of state including partial ionization through Saha's approximation, the pressure of gas and radiation as well as the equations for degenerate electrons.

As next step, an empirical effective temperature calibration for main sequence (luminosity class V to III) stars in terms of $(g_1-y)_0$ is established. This is most important for studying very distant galactic open clusters and extragalactic systems for which, in general, no photometric data within a standard system are available. The absolute magnitudes and thus luminosities can be easily estimated via y and the appropriate isochrones.

For the empirical temperature calibration, a homogeneous sample of bright ($V < 7$ mag), apparently normal type objects, from Section 10 was taken. In total, 225 stars were used to derive effective temperatures within the Strömgren $uvby\beta$, Geneva 7-colour and Johnson UBV systems which were then applied to establish a calibration in terms of $(g_1 - y)_0$. The final calibration is valid for effective temperatures between 33 000 and 10 000 K and yields a statistical mean error of 238 K for the whole spectral range.

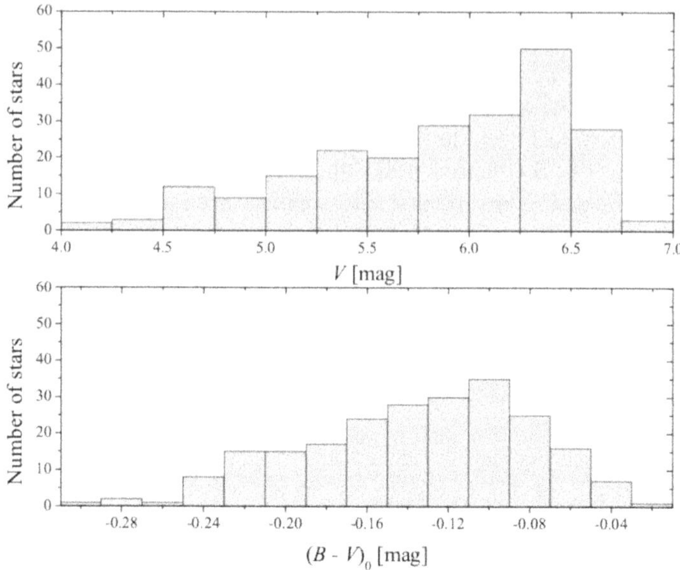

Figure 25: The distribution of Johnson V and $(B - V)_0$, see Sect. 12.1.2, for the sample of 225 main sequence B-type objects.

12.1.1 Sample of program stars

The "normal" B-type objects were selected fulfilling the following criteria:

- classification as B-type, luminosity class V to III
- no significantly deviating Δa values, i.e. ± 10 mmag
- not listed in the catalogue of Ap and Am stars by Renson & Manfroid (2009, RM09)
- available data within the Johnson, Strömgren, Geneva, and Δa photometric system

Binary systems of all kinds and high $v \sin i$ stars were a-priori not excluded. The Johnson, Geneva, and Strömgren colours were taken from the General Catalogue of Photometric Data (Mermilliod et al., 1997, GCPD). The Δa photometry is from Section 10.

The following stars have inconsistent photometric measurements and were therefore excluded: HR 345, HR 1375, HR 1617, HR 2870, HR 3470, HR 8854, and HR 8887. From the available data, it was not possible to decide if measurement errors, a wrong identification or a binary nature causes these discrepancies.

The final list comprises 225 objects that satisfy the criteria. The complete list of program stars is only available in electronic form. This table includes the identification of objects, the $(g_1 - y)_0$, $(B-V)_0$, X and $(u-b)$ values, V magnitudes, effective temperature with the corresponding errors (Sect. 12.1.3), $v \sin i$ values, and spectral types, respectively.

The distribution of Johnson V and $(B-V)_0$, see Sect. 12.1.2 for the estimation of the reddening, for the sample is shown in Figure 25. The peak of the $(B-V)_0$ values is at about -0.10 mag (B8, Table 13) which reflects the coincidence that the CP stars which were measured within the Δa photometric system peak also at this effective temperature (Schneider, 1993) and the normal type objects, used in this investigation, served as standard stars.

12.1.2 The estimation of the reddening

The reddening for B-type stars within the solar neighbourhood is, in general, estimated using photometric calibrations in the Strömgren $uvby\beta$ (Crawford, 1978) and the Q parameter within the Johnson UBV system (Johnson, 1958). These methods are only based on photometric indices and do not take into account any distance estimates via parallax measurements.

The reddening in the Strömgren $uvby\beta$ photometric system is based on the comparison of the reddened $(b-y)$ and c_1 with the unreddened $(u-b)$ and β indices (Crawford, 1978). The procedure of the Q method is straightforward and has been described in much details by Johnson (1958). Here are the basic correlations from this reference:

$$Q = (U-B) - 0.72 \cdot (B-V) \quad (12)$$

$$E(B-V) = (B-V) - 0.332 \cdot Q \quad (13)$$

Throughout these sections, the following relation is used: $A_V = 3.1E(B-V) = 4.3E(b-y)$. Figure 26 shows the comparison of the derived A_V values for both methods. The agreement is very good indicating that photometric measurements within the Johnson and Strömgren systems are intrinsically consistent. Otherwise, a severe deviation from the linear correlation would occur. The distribution of the adopted values (mean of both methods) shows that most of the program stars have an absorption which is below 0.1 mag.

However, the interstellar reddening was also derived using the model proposed by Chen et al. (1998) who combined galactic reddening maps, which are derived from open clusters as well as from galactic field stars

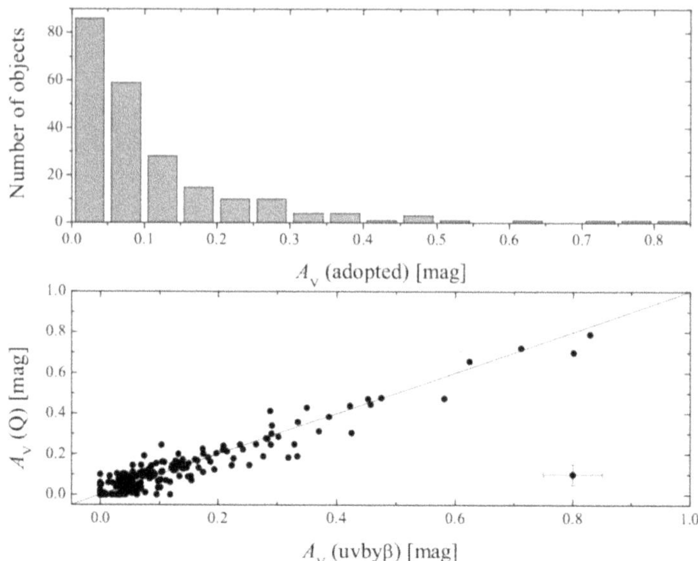

Figure 26: The correlation of the absorption A_V estimated from the Strömgren $uvby\beta$ and the Q method (upper panel). The statistical error for both methods is indicated with the symbol in the lower left corner. The distribution of the adopted A_V values is shown in the lower panel.

with published empirical reddening laws from the literature. As input parameters, the galactic coordinates and the distance from the Sun (i.e. derived from Hipparcos parallax measurements) are needed. The error of the latter severely influences the error of the derived reddening. For program stars with distance errors smaller than 15%, a very good agreement with the results of the photometric calibrations has been found.

12.1.3 The calibration of the effective temperature

To reach the final goal, an empirical effective temperature calibration of $(g_1 - y)_0$ for B-type stars, this astrophysical parameter was calibrated for the sample using the published calibrations in the Geneva, Strömgren, and Johnson photometric systems. Those calibrations are, in general, derived independently from each other which allows to detect possible inconsistencies due to, for example, spectroscopic binaries. The applied calibrations are now discussed in more details.

Geneva system: detailed calibrations were published by Cramer (1984, 1999); Künzli et al. (1997). They are all based on the reddening free parameters X and Y which are valid for spectral types hotter than approximately A0. The results are therefore independent of the estimation of A_V for the program stars.
Strömgren system: the most recent and widely used reference is Napiwotzki et al. (1993). For stars hotter than 11000 K, the unreddened $[u-b]$ and for cooler B-type objects, the a_0 index are used to calibrate the effective temperature. The latter is not reddening free.
Johnson system: first of all, the Q values for luminosity class III and V objects were calculated according to the Tables listed by Schmidt-Kaler (1982). The $(B-V)_0$ values for those luminosity classes were transformed into effective temperatures using the results by Code et al. (1976, Table 7). As final correlation a polynomial fit for dependence of the effective temperature on Q is derived as

$$\log T_{\rm eff} = 3.983(5) - 0.31(3) \cdot Q + 0.33(3) \cdot Q^2 \tag{14}$$

which is valid for all B-type, luminosity class III to V objects. This relation has to be treated as an averaged statistical result. If one uses the individual measurements from Code et al. (1976) for the ten stars that are in the relevant effective temperature and luminosity regime (Sect. 12.1.1), the following relation is derived:

$$\log T_{\rm eff} = 3.994(29) - 0.25(15) \cdot Q + 0.38(15) \cdot Q^2 \tag{15}$$

A comparison of these two relations gives $\Delta T_{eff} = +141(147)$ K.

The individual effective temperature values within each photometric system were first checked for their intrinsic consistency and then averaged. These final values together with the standard deviations of the means are listed in the electronically available table. No outliers in any photometric system were detected.

12.1.4 The calibration procedure for the Δa photometric system

The starting point is one set of $(g_1 - y)$ measurements, for example, of an open cluster. Here one faces the first problem because the zero points of data sets from different instruments and filter systems may vary. An overview of several used Δa filters is listed in Section 10.

First of all, the reddening coefficient k of the relation $E(g_1 - y) = k \cdot E(B-V)$ was estimated which can then be transformed to other photometric systems. The correlation coefficients for $(B-V)$ versus $(g_1 - y)$

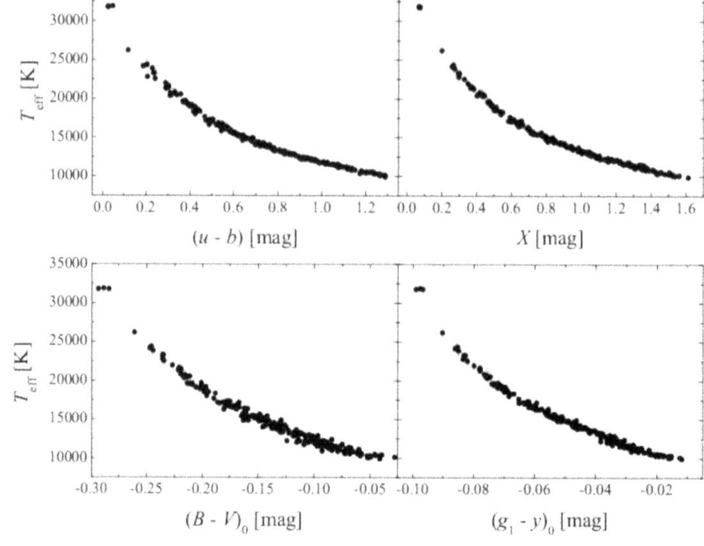

Figure 27: Mean relation between the effective temperature and $(u-b)$, X, $(B-V)_0$ as well as $(g_1-y)_0$ for B-type, luminosity class V to III objects.

are derived and then applied to $(B-V)_0$ using the estimated reddening for each individual star as well as (g_1-y) to derive $(g_1-y)_0$ and thus $E(g_1-y)$. The final value, $k=0.39(2)$ is in excellent agreement with those values derived from open cluster data.

As next steps, the individual indices are dereddened and the linear correlations between those parameters are established. A "standard (g_1-y) system" is defined which is set to $(B-V)_0=(b-y)_0=(g_1-y)_0=0$ as following:

$$(B-V)_0 = -1.413(7) + 2.82(1) \cdot (g_1-y)_0 \tag{16}$$

$$(b-y)_0 = -0.664(5) + 1.333(9) \cdot (g_1-y)_0 \tag{17}$$

This system was also kept for the the calibration of the A-type to early F-type stars (Sect. 12.2).

Figure 27 shows the relation between the effective temperature and the different temperature sensitive indices for the four investigated photometric systems. The temperature range from approximately 33 000 to 10 000 K which covers the main sequence B-type stars. However, there are only four stars with temperatures hotter than 25 000 K (B1.5 V). The temperature

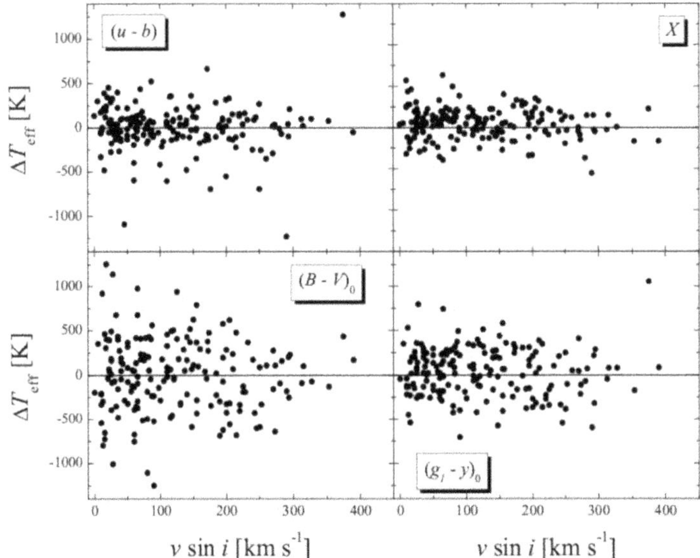

Figure 28: No systematic correlation of $\Delta T_{\rm eff}$ and the $v \sin i$ values were found for $(u-b)$, X, $(B-V)_0$ as well as $(g_1-y)_0$. The well-known effect of the rotational velocity on the photometric indices is therefore less significant than the precision of the method itself.

relations were checked with and without those four data points and no significant differences were found. The final calibrations are given as:

$$\begin{aligned}
\log T_{\rm eff} = & + 4.520(3) - 0.754(15)\cdot(u-b) + 0.413(2)\cdot(u-b)^2 - \\
& - 0.108(11)\cdot(u-b)^3 && (18)\\
= & + 4.546(3) - 0.698(13)\cdot X + 0.367(17)\cdot X^2 - \\
& - 0.092(6)\cdot X^3 && (19)\\
= & + 3.956(5) - 1.04(7)\cdot(B-V)_0 + 2.89(24)\cdot(B-V)_0^2 && (20)\\
= & + 3.909(7) - 6.47(48)\cdot(g_1-y)_0 - 47(9)\cdot(g_1-y)_0^2 - \\
& - 425(60)\cdot(g_1-y)_0^3 && (21)
\end{aligned}$$

with the mean of the errors for the whole sample of $\Delta T_{\rm eff}[(u-b), X, (B-V)_0, (g_1-y)_0] = [157{,}146{,}333{,}238\,\mathrm{K}]$. These are statistical errors and should be treated as such. The mean relations between the effective temperature and $(B-V)_0$, $(u-b)$, X as well as $(g_1-y)_0$ depending on spectral types are listed in Table 13. The agreement with standard values for those indices from the literature (Code et al., 1976; Crawford, 1978; Cramer, 1999) is excellent.

It is known that high rotational velocities can alter the photometric indices significantly (Collins et al., 1991). The break-up velocity ranges from about 540 to 400 $\mathrm{km s^{-1}}$ for B0 to B9 stars (Townsend et al., 2004). However, the inclination i is the crucial point for the comparison of these models with observations. One only has to recall the case of the prototype A0 star Vega that has a very low $v \sin i$ of 22 $\mathrm{km s^{-1}}$ but an equatorial velocity of 160 $\mathrm{km s^{-1}}$ (Hill et al., 2010). The papers by Glebocki & Stawikowski (2000); Abt et al. (2002) were searched for available $v \sin i$ values for the program stars, and averaged individual measurements from the different sources. In total, there are measurements for 175 stars of the sample available with the highest value of 390 $\mathrm{km s^{-1}}$ for HR 3502. Figure 28 shows the diagrams of $\Delta T_{\mathrm{eff}} = T_{\mathrm{eff}}(\mathrm{orig}) - T_{\mathrm{eff}}(\mathrm{calib})$ versus $v \sin i$ for the different indices. There is no obvious correlation between these two parameters in any of the diagrams evident. The effect of the rotational velocity is therefore less significant than the precision of the method itself. Or in other words, the well-known alteration of stellar colours caused by high rotational velocities (Collins et al., 1991) cannot be distinguished from the overall statistical errors resulting from the calibration process.

The overall procedure of deriving the effective temperatures for main sequence B-type objects in the Δa photometric system should be as following:

- Estimate the reddening, for example via isochrones for open cluster members
- $E(g_1 - y) = 0.39 \cdot E(B - V)$
- Apply the reddening correction for all individual indices
- Transform the $(g_1 - y)_0$ via the standard relations
- Check the intrinsic consistency of all available measurements according to the spectral type - effective temperature - photometric indices correlation
- Apply the effective temperature calibration

The estimated effective temperature from the Δa system can in addition compared with calibrated values from other photometric indices.

12.2 The A-type and mid F-type stars

Here, the A-type to mid F-type objects are investigated that exhibit an increase line blanketing and luminosity effects without the availability of

Table 13: Mean relation between the effective temperature and $(B-V)_0$, $(u-b)$, X as well as $(g_1-y)_0$ for B-type, luminosity class V to III objects.

Spec.	$T_{\rm eff}$	$(B-V)_0$	$(u-b)$	X	$(g_1-y)_0$
B0	32000	−0.292	0.020	0.060	−0.102
B1	26000	−0.257	0.151	0.210	−0.090
B2	23000	−0.236	0.239	0.310	−0.082
B3	18000	−0.189	0.448	0.557	−0.065
B5	16000	−0.164	0.570	0.705	−0.056
B6	14500	−0.142	0.688	0.850	−0.049
B7	13500	−0.125	0.785	0.970	−0.043
B8	12500	−0.105	0.901	1.113	−0.036
B9	10800	−0.064	1.155	1.411	−0.021
(A0)	9800	−0.031	1.338	1.603	−0.011

any a-priori reddening-free parameter. Only Strömgren β does, in general, not depend on the extinction. However, it is sensitive not only to the effective temperature alone but also to the luminosity (Gerbaldi et al., 1999). Based on the methods used in Sect. 12.1, the following parameters were used for this purpose: $(b-y)_0$, $(B2-V1)_0$, and $(B-V)_0$. The scatter of the derived effective temperatures for spectral types between A0 and A3 is not larger than that for later type stars because the sample of bright galactic-field stars is almost free of reddening.

Applying the same selection criteria as for the B-type stars (Sect. 12.1.1) yields 282 luminosity class V to III A-type to mid F-type objects. The derived mean errors for the effective temperature calibrations within all photometric systems are smaller than those for the B-type stars.

12.2.1 Sample of program stars

The selection criteria and the sources of photometry for the A-type to mid F-type objects are exactly the same as listed in Section 12.1.1.

The final list includes 282 objects. This table includes the identification of objects, the complete Strömgren $uvby\beta$, $(g_1-y)_0$, $(B-V)_0$, and $(B2-V1)_0$ values, V magnitudes, $E(b-y)$ values, effective temperature with the corresponding errors, $v\sin i$ values, and spectral types, respectively.

The distribution of Johnson V and $(B-V)_0$ for the sample is shown in Figure 29. The estimation of the reddening is explained in more detail in Section 12.2.2. The distribution of V is comparable to the one for the B-type sample peaking at $V=6.25$, mag, whereas the one for $(B-V)_0$

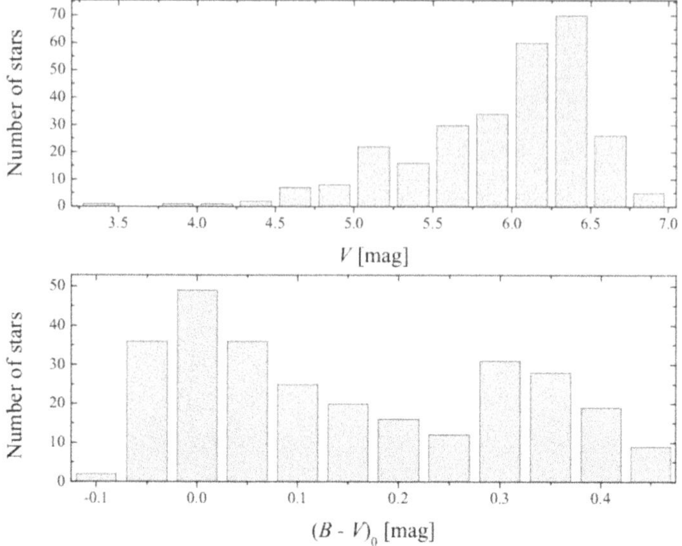

Figure 29: The distribution of Johnson $(B-V)_0$ (lower panel) and V (upper panel) for the used sample of 282 main sequence A- to mid F-type objects. The distribution of $(B-V)_0$ shows two maxima at spectral types of about A0 and F0.

exhibits two maxima at spectral types of about A0 (+0.0 mag) and F0 (+0.3 mag, see also Table 14). However, there is a statistically significant number of objects in the complete investigated spectral range.

12.2.2 The estimation of the reddening

The sample comprises stars later than a spectral type of A0. A typical A0 main sequence object has a mean absolute magnitude of about +0 mag. A visual magnitude of 7 mag (Fig. 29) then corresponds to a distance of 250 pc, for which the reddening can be almost neglected in all directions (Neckel et al., 1980).

The following photometric calibrations in the Strömgren $uvby\beta$ system were used to estimate the reddening according to the valid spectral range that is estimated by the standard relations of the different indices:

- A0 – A3: Crawford (1978); Hilditch et al. (1983)
- A3 – F0: Crawford (1979); Domingo & Figueras (1999)

- later than F0: Crawford (1975); Schuster & Nissen (1989)

The calibrations are not very reliable for stars with spectral types between A0 and A3 (Gerbaldi et al., 1999), mainly because for these stars, the reddening-free parameter β is no longer a temperature indicator alone but is also sensitive to the luminosity, therefore the standard relation of $(b-y)_0$ versus m_0 as listed by Hilditch et al. (1983) and their method of calibration were used. Only some objects that fall outside the given relation were dereddened using the calibration by Crawford (1978).

Furthermore, the interstellar extinction model by Chen et al. (1998) used to derive the reddening for all program stars as described in Section 12.1.2. The values from the calibration of the Strömgren $uvby\beta$ and the model by Chen et al. (1998) agree very closely. As expected, all objects have a calibrated total absorption A_V of less than 0.35 mag with 233 (82% of the complete sample) stars even lower than 0.05 mag. Taking the following relations into account:

$$A_V = 3.1E(B-V) = 4.3E(b-y) = 4.95E(B2-V1) = 7.95E(g_1-y), \quad (22)$$

the effect of the reddening on the calibration for the sample can be neglected and does not introduce a significant error source.

12.2.3 The calibration of the effective temperature

As for the B-type sample, the first step was to derive the effective temperature for each individual star within the Geneva, Strömgren, and Johnson photometric systems. For this purpose the reddening relations as listed in Eq. 22 was used to calculate the unreddened indices that are necessary to make the proper calibration as listed.

Geneva system: the calibration by Künzli et al. (1997) is the most recent for the investigated spectral range. For intermediate stars hotter than 8500 K, they used the parameters pT and pG, which are linear combinations of the seven Geneva colors. Those two indices can be dereddened with the following relations: $pT_0 = pT - E(B2 - V1)$ and $pG_0 = pG - 1.1E(B2 - V1)$. For cooler objects, the grids of m_2 and d versus $(B2 - V1)_0$ served as a calibration tablet. The definition of these indices are listed in Golay (1994).

Strömgren system: Napiwotzki et al. (1993) investigated several calibrations based on a_0 and r^* for hotter, as well as β, and c_0 for cooler objects yielding a rather unsatisfactory result. Finally, they established a $T_{\rm eff}$ versus $(b - y)_0$ relation (Eq. 10 therein), which was applied to sample here.

Johnson system: the semi-empirical $T_{\rm eff}$ versus $(B - V)_0$ relation as listed in Gray (2008, Eq. 15.14) was used. It is based on synthetic colours

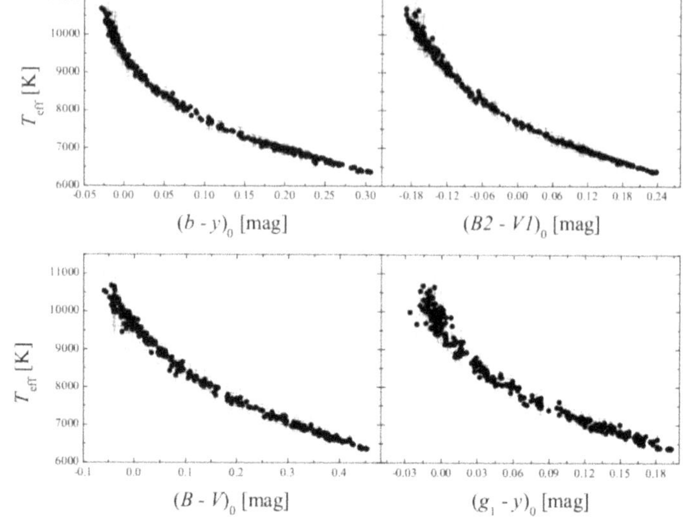

Figure 30: Mean relation between the effective temperature and $(b-y)_0$, $(B2-V1)_0$, $(B-V)_0$, as well as $(g_1-y)_0$ for A-type to mid F-type, luminosity class V to III objects.

from theoretical stellar atmospheres that are normalized to observations of spectroscopic binary systems, as well as bright stars. It is given as

$$\begin{aligned}\log T_{\mathrm{eff}} = &+ 3.988 - 0.881(B-V)_0 + 2.142(B-V)_0^2 - 3.614(B-V)_0^3 + \\ &+ 3.2637(B-V)_0^4 - 1.4727(B-V)_0^5 + 0.2600(B-V)_0^6 \quad (23)\end{aligned}$$

and is valid for all A-type to mid F-type, luminosity class III to V objects. It superseded the relation listed in Code et al. (1976).

The individual effective temperature values for the three photometric systems were first tested for their intrinsic consistency and then averaged. No statistical significant outliers in any photometric system were detected.

12.2.4 The result for the Δa photometric system

The relation between the mean effective temperature and the different temperature sensitive indices for the four investigated photometric systems is shown in Figure 30. The errors are constant over the complete spectral range. There is a larger scatter only for the Geneva $(B2-V1)_0$ index at effective temperatures hotter than 9000 K. This region was checked with

Table 14: Mean relation between the effective temperature and $(B-V)_0$, $(b-y)_0$, $(B2-V1)_0$, as well as $(g_1-y)_0$ for A-type to mid F-type, luminosity class V to III objects. Only spectral types according to the Yerkes system (Jaschek & Jaschek, 1990) are listed.

Spec.	$T_{\rm eff}$	$(B-V)_0$	$(b-y)_0$	$(B2-V1)_0$	$(g_1-y)_0$
A0	10000	−0.025	−0.015	−0.160	−0.010
A2	8750	+0.067	+0.031	−0.093	+0.025
A3	8300	+0.112	+0.055	−0.060	+0.042
A5	7900	+0.161	+0.082	−0.024	+0.062
A7	7500	+0.221	+0.120	+0.022	+0.087
F0	7050	+0.304	+0.185	+0.087	+0.125
F2	6750	+0.368	+0.238	+0.141	+0.155
F3	6650	+0.390	+0.254	+0.159	+0.165
F5	6450	+0.433	+0.284	+0.196	+0.184

the grids of pT_0 and pG_0 originally used. It is exactly where both indices become zero for the Main Sequence and small shifts result in larger uncertainties. It therefore seems to be an intrinsic numerical problem of the grids themselves; however the overall statistical error of $(B2-V1)_0$ is satisfying. The final calibrations parameterized as third-degree polynomials are:

$$\begin{aligned}\log T_{\rm eff} = & + \ 3.9793(4) - 1.34(2)\cdot(b-y)_0 + 4.90(15)\cdot(b-y)_0^2 - \\ & - \ 8.06(38)\cdot(b-y)_0^3 \qquad (24)\\ = & + \ 3.8853(4) - 0.459(5)\cdot(B2-V1)_0 + 1.02(2)\cdot(B2-V1)_0^2 - \\ & - \ 2.36(16)\cdot(B2-V1)_0^3 \qquad (25)\\ = & + \ 3.9825(4) - 0.670(10)\cdot(B-V)_0 + 1.05(7)\cdot(B-V)_0^2 - \\ & - \ 0.98(12)\cdot(B-V)_0^3 \qquad (26)\\ = & + \ 3.9817(8) - 1.79(5)\cdot(g_1-y)_0 + 8.14(77)\cdot(g_1-y)_0^2 - \\ & - \ 19(3)\cdot(g_1-y)_0^3 \qquad (27)\end{aligned}$$

with the mean of the errors for the whole sample of $\Delta T_{\rm eff}[(b-y)_0,(B2-V1)_0,(B-V)_0,(g_1-y)_0]$ = [66,100,71,134 K]. These are statistical errors for the complete sample. The errors for individual stars are, of course, larger than that. Note that the error for the Johnson $(B-V)_0$ calibration is surprisingly small. It shows the overall robustness of broad-band indices against metallicity and luminosity effects, which makes them superior for statistical analysis of larger samples over a wide spectral range. However, it has to be emphasized that it is a-priori not possible to apply dereddening methods in the Johnson UBV system for stars cooler than B9 (Johnson, 1958).

The statistical error for $(g_1 - y)_0$, 134 K, is significantly smaller than the one for the hotter stars (238 K). But one has to keep in mind that the absolute errors are over the complete spectral range from early B-type to F-type constant, but only the relative one, as listed, is decreasing.

Table 14 lists the mean relations between the effective temperature and $(B - V)_0$, $(b - y)_0$, $(B2 - V1)_0$, as well as $(g_1 - y)_0$, depending on spectral types. Only standard spectral types of the Yerkes system (Jaschek & Jaschek, 1990) are given. The $(B - V)_0$ and $(g_1 - y)_0$ values for A0 are, within the error limits, identical to those derived for the B-type sample, which guarantees an intrinsic consistent calibration from B0 to F5 for the Δa photometric system.

The effect of high $v \sin i$ objects on the derived calibrations was checked in the same manner as described in Section 12.1.4. This might be an issue for early A-type objects only because the rotational velocities decrease significantly for cooler objects (Zorec & Royer, 2012). Again, no statistical significant effect of the rotational velocity, which can be distinguished from other error sources, was found on the precision of the calibration.

The detailed calibration procedure for determining the effective temperature within the Δa photometric system is given in Section 12.1.4. Here, only a short overview of the main method is given. After the estimation of the reddening $E(g_1 - y)$, the dereddened standard $(g_1 - y)_0$ has to be calculated and the calibration for the appropriate spectral region applied. As a final check, a comparison with the values for other photometric systems, if available, should be performed.

13 Theoretical isochrones for the Δa photometric system

Theoretical isochrones for the photometric Δa system were calculated to derive astrophysical parameters such as the age, reddening, and distance modulus for star clusters. Using the stellar evolutionary models by Claret (1995), a grid of isochrones with different initial chemical compositions for the Δa system was generated. The data of 23 open clusters (Sect. 17) were used to fit these isochrones with astrophysical parameters (age, reddening, and distance modulus) from the literature. As an additional test, isochrones with the same parameters for Johnson UBV data of these open clusters were also considered. The fits show a good agreement between the observations and the theoretical grid. The accuracy of fitting isochrones

Table 15: 23 published open clusters with Δa CCD photometry taken from Section 17. The metallicity for all clusters is (within the errors) solar according to DAML02 (Dias et al., 2002).

Cluster	$\log t$	$m_V - M_V$	$E(B-V)$	N_{Stars}	Cluster	$\log t$	$m_V - M_V$	$E(B-V)$	N_{Stars}
Collinder 272	7.20	13.06	0.47	111	NGC 6134	8.97	11.07	0.40	102
Lyngå 14	6.71	14.29	1.48 (var)	53	NGC 6192	7.95	13.00	0.68	98
Melotte 105	8.30	12.70	0.36	114	NGC 6204	7.60	11.50	0.43	319
NGC 2099	8.50	11.67	0.30	41	NGC 6208	9.00	10.54	0.18	41
NGC 2169	7.70	10.57	0.12	13	NGC 6250	6.50	11.18	0.38	48
NGC 2439	7.20	14.23	0.41	113	NGC 6396	7.20	13.34	0.96 (var)	105
NGC 2489	8.45	12.00	0.40	59	NGC 6451	8.13	13.74	0.67	146
NGC 2567	8.43	11.40	0.13	50	NGC 6611	6.90	13.72	0.86 (var)	79
NGC 2658	9.10	11.67	0.04	55	NGC 6705	8.30	12.73	0.43	312
NGC 3114	8.10	10.02	0.07	271	NGC 6756	8.10	14.67	0.80	65
NGC 3960	8.80	12.74	0.30	93	Pismis 20	6.60	15.30	1.24 (var)	238
NGC 5281	7.10	11.35	0.26	30					

to Δa data without the knowledge of the cluster parameters is between 5 and 15 %.

13.1 Models, data selection, and isochrone fitting

The stellar models used to calculate the isochrones have been described in more detail by Claret (1995). Here a short overview of the input physics is given. The chemical composition is (X, Z) = (0.70, 0.02) though different combinations of X and Z can also be used following the metallicity indicator of each cluster. The equation of state takes into account partial ionization through Saha's approximation, the pressure of gas and radiation as well as the equations for degenerate electrons. For less massive stars, a special treatment of the equation of state was adopted through the CEFF package (Däppen, 2000). Radiative opacities were computed with the OPAL code. The mixing length theory was used to describe convection and moderate overshooting with $\alpha_{\rm ov} = 0.2$ was considered for convective cores. The models take into account mass loss during the main sequence as well as during the red giant phase.

The filter transmission functions of g_1 and y are the same as those used in Section 14. Synthetic colour calculations were performed using the properties of the mentioned filters, in order to establish the connection between observed and theoretical quantities. A similar procedure as described by Castelli (1999) was adopted. The original ATLAS9 models Kurucz (1993), with a microturbulence velocity of 2 kms^{-1}, $-5.0 \leq \log\,[M/H] \leq +1$, and mixing-length parameter $\alpha = 1.25$ were the basic tools to derive the synthetic colours. For $T_{\rm eff} \leq 8500$ K, the calculations listed in Sect. 14 were adopted due to problems detected in the convection model, precisely in the specific intensities (Claret, 1995). These models were computed adopting

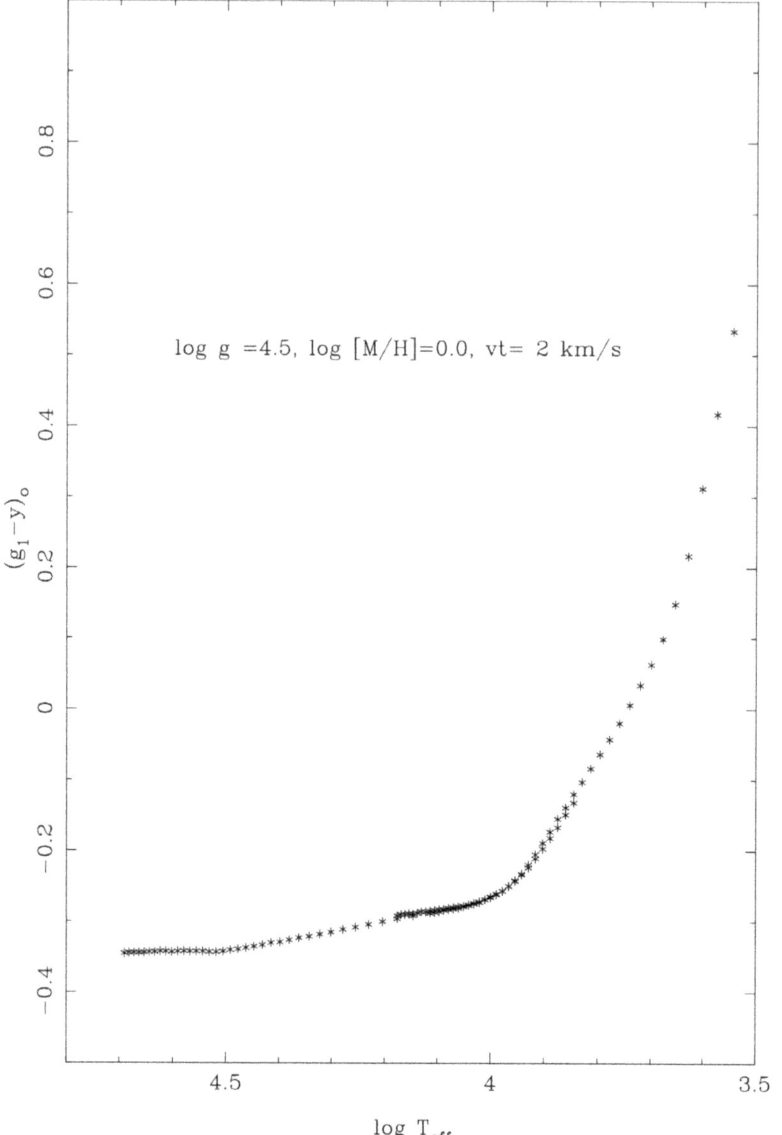

Figure 31: Theoretical relationship between $(g_1 - y)_0$ and the effective temperature for models with solar abundance, $\mathrm{vt} = 2\,\mathrm{kms}^{-1}$, $l/H_p = 1.25$ and $\log g = 4.5$ dex. For models with $T_{\rm eff} \leq 8500$ K, an alternative theory of convection was adopted.

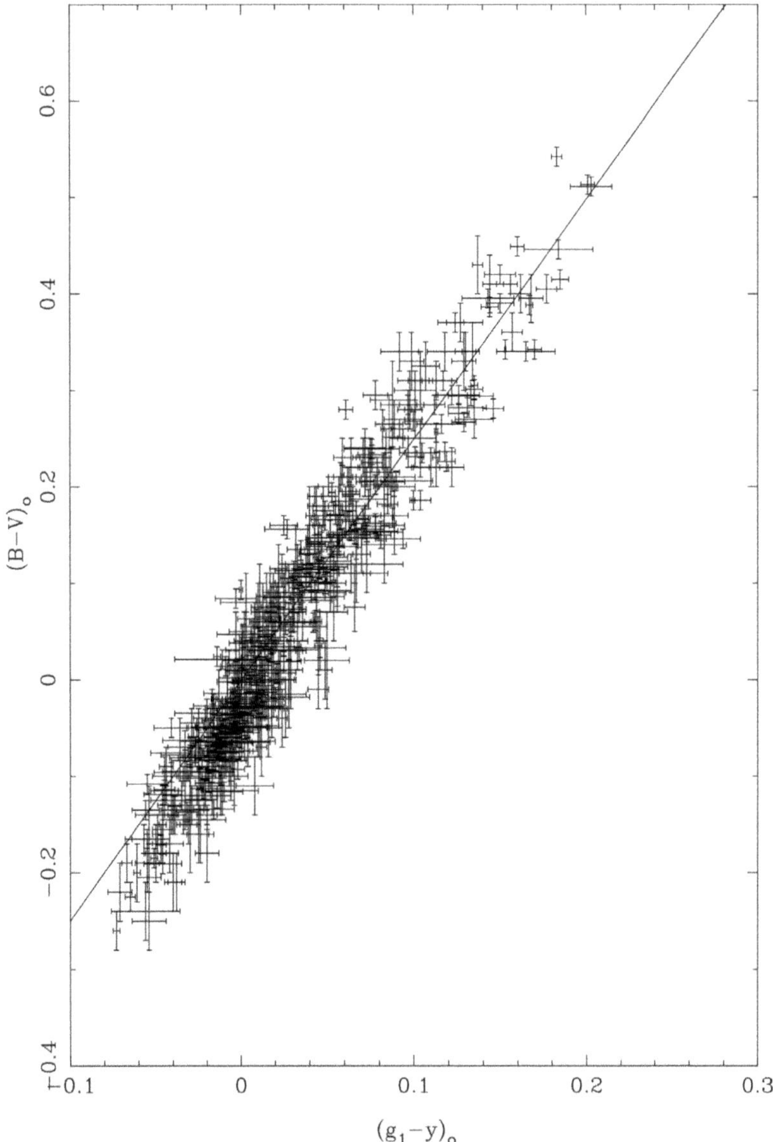

Figure 32: The empirical $(g_1 - y)_0$ and $(B - V)_0$ data and the respective theoretical predictions (continuous line). The theoretical value was obtained by averaging the whole grid of stellar atmosphere models.

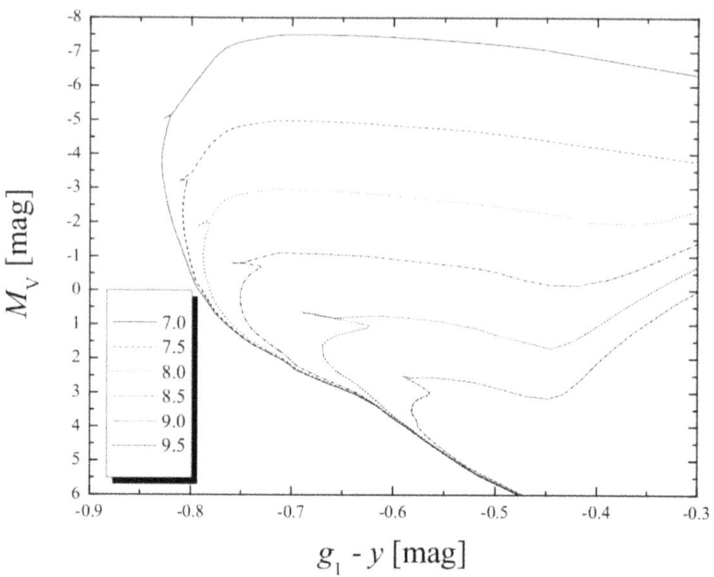

Figure 33: Theoretical isochrones for different ages and solar abundance for the evolutionary models described in Section 13.1.

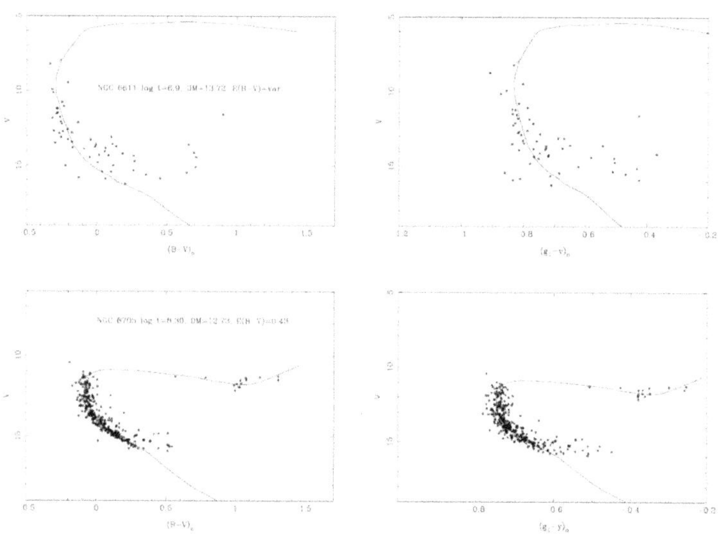

Figure 34: Isochrones for NGC 6611 and NGC 6705 with the parameters listed in Table 15.

the Canuto & Mazzitelli (1991) prescription. The implementation of this alternative theory of transport of energy by convection does not significantly affect the hotter models, as expected. In fact, the differences in the respective colour indices are of the order of 1 to 3 mmag, as described and calculated in Section 14. As an example of synthetic colour calculation, Fig. 31 shows the dependence of $(g_1-y)_0$ with the effective temperature for models with $\log g=4.5$ dex. Similar results are obtained for different values of $\log g$, microturbulent velocities and/or metallicity. Figure 32 shows a comparison between empirical data and theoretical colour indices. The continuous line indicates the average of all theoretical models (in $\log g$). The zero-point was corrected by adding 0.31 to the theoretical (g_1-y) values. The final isochrones, considering the adequate values of $\log g$ for each class of luminosity, are shown in Fig. 33 for different ages from $\log t=7.0$ to 9.5, respectively.

In Sect. 14 it is shown that metallicities different from solar values shift the normality line by about 3 to 6 mmag for $[Z] = \pm 0.5$ dex. Such an effect is, in general, a factor of two smaller than the intrinsic measurement errors and not detectable. Nevertheless, the DAML02 database Dias et al. (2002) was searched through for available metallicities of the open clusters listed in Table 15. For none of the investigated clusters, a value significant different than solar was found within the given error. Therefore it is expected that all programme clusters have solar metallicity.

The published Δa CCD photometry together with the Johnson UBV ones of 23 open clusters (Sect. 17) was taken to test these isochrones. These open clusters have widely different ages and reddening which makes them excellent test cases (Table 15).

The isochrone fitting was performed in two steps. First, the (g_1-y) values were dereddened according to $E(g_1-y) = 0.4 \cdot E(B-V)$ and an individual isochrone was calculated according to the ages (taken from the literature) listed in Table 15. The data were then plotted with the appropriate distance modulus. In order to test the parameters from the literature, the same procedure was performed for Johnson UBV colours. Figure 34 shows the examples of NGC 6611 and NGC 6705 for both photometric systems, respectively. The isochrones fit, in general, the data very well.

As second step, the Δa data were fitted to the isochrones without an a-priori knowledge of the reddening, age and distance modulus. Here the same problems and error sources are evident as for the classical photometric systems since no colour-colour diagram is available. Nevertheless, it was possible to reproduce the parameters from the literature with an accuracy between 5 and 15% depending on the age, available giant members and the presence of differential reddening.

14 Synthetic Δa photometry

A synthetic photometric system was developed which can be used to explore the capability of model atmospheres with individual element abundances. This can be used to predict photometric Δa magnitudes which measure the extent of the flux depression around 5200Å found in different types of CP stars (Sect. 6).

One of the main conclusions drawn from the explanation of the 5200Å flux depression (Sect. 8.1) has been the necessity to build specific model stellar atmospheres for CP stars using state-of-the-art opacity data. The increase in available computer power and advances in computational algorithms during the last two decades have now brought this problem into the realm of workstations and personal computers. Moreover, stellar atmosphere modelling can take advantage of data bases for atomic line transitions devoted to stellar atmosphere applications such as Kurucz (1992) and the VALD project (Kupka et al., 1999). Among the current projects for the computation of model atmospheres there are several which can calculate models with individual abundances on workstations or personal computers. Two of them are based on an opacity sampling approach. ATLAS12 by Kurucz (2005), for which applications were presented by Castelli & Kurucz (1994), is particularly suitable for B- to K-type stars at or close to the main sequence. The MARCS project in Uppsala (Bengt Gustafsson and his group) is aimed at the cool part of the HRD and can produce model atmospheres for stars with spectral types later than A0 (Gustafsson et al., 2008). However, for the computation of small grids of model atmospheres with individual abundances, which are required when varying T_{eff} or $\log g$ during the initial analysis of a single star or several sufficiently similar stars, the opacity distribution function (ODF) approach remains preferable due to its higher computational efficiency. Piskunov & Kupka (2001) have presented a new software toolkit based on this approach using a modified version of the ATLAS9 code of Kurucz (1993). It is suitable for spectral types from early B-type to early F-type stars at or close to the main sequence.

14.1 The normality line

First, the observed dependency of the a index as a function of various colour indices sensitive to the effective temperature of stars as well as its average

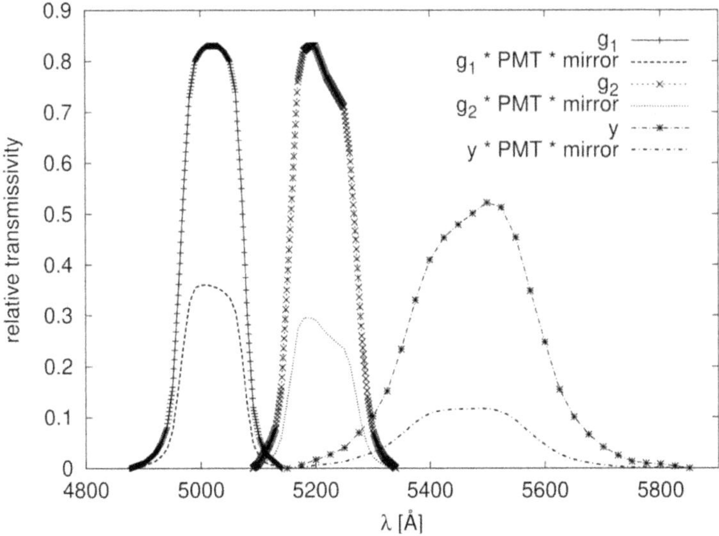

Figure 35: Filter transmission functions and effective transmission of the synthetic photometric system after convolution with the profiles of the response function of a 1P21 RCA photomultiplier tube (PMT) and a typical mirror reflection efficiency function. The latter two are taken from the UVBY.FOR programme of Kurucz (1993).

scatter expected from surface gravity variations within the main sequence band was investigated. The behaviour of the so-called "normality line" of Δa systems used in photometric observations of CP stars (Sect. 3) is well reproduced. The metallicity dependence of the normality line of the Δa system was computed for several grids of model atmospheres where the abundances of elements heavier than He had been scaled ± 0.5 dex with respect to the solar value. A lowering of Δa magnitudes for CP stars within the Magellanic Clouds by ~ -3 mmag relative to those in the solar neighbourhood assuming an average metallicity of [Fe/H] $= -0.5$ dex was estimated. Using these results on the metallicity bias of the Δa system one finds the observational systems in use suitable to identify CP stars in other galaxies or distant regions of our own galaxy and capable to provide data samples on a statistically meaningful basis.

14.1.1 The synthetic Δa filter system

Relyea & Kurucz (1978) have discussed in detail how to calculate synthetic colours from model atmosphere fluxes computed with the ATLAS code. The main idea is to convolve the emergent surface fluxes predicted from model atmospheres with several functions representing filter transmission, relative absorption of all other devices in the optical path (including telescope mirrors), and detector sensitivity.

Figure 35 shows the response functions of the applied synthetic photometric system. The steep decay of the response function of the 1P21 RCA photomultiplier from 58% at 5000Å to 29% at 5500Å and a mere 10% at 6000Å explains the smaller effective sensitivity of the system in the y-band relative to the g_1-band in comparison with the transmission functions of the filters themselves. Here it has to be noted that the transmission functions for the g_1 and g_2 filters were taken from calibration measurements of one of the filter sets used in earlier observations (System "2" in Maitzen & Vogt, 1983), while the y filter transmission (difference less than 1 % to that of Maitzen & Vogt, 1983) and the mirror reflection efficiency and detector response functions were taken from the UVBY code of Kurucz (1993). Maitzen & Vogt (1983) used the response function of an EMI 6256 photomultiplier for their calibrations which has essentially the same shape but a different absolute sensitivity as the 1P21 one in the relevant wavelength region. This offset in the absolute sensitivity does not affect the synthetic magnitudes since only differences between individual filters were used.

Only photoelectric Δa measurements (Sect. 10) were used for the comparisons here. However, there are also a lot of CCD Δa measurements available by now (cf. Sect. 18). A classical photomultiplier is almost insensitive at approximately 6000Å whereas a CCD is very sensitive in the red region. This implies an almost linear increase of the response function from g_1 to y. In the new CCD system, the FWHM of the g_1 and y filters are 222Å and 120Å, respectively. This guarantees that the total flux of each filter after the convolution with the response function is comparable with that one of the "old" system.

14.1.2 Calibration relations

The overall success of the ATLAS9 model atmospheres of Kurucz (1993) to reproduce photometric colours and spectrophotometric fluxes of standard stars of spectral types B and A (Castelli & Kurucz, 1994; Smalley & Dworetsky, 1995; Castelli et al., 1997) and also for some of the CP stars (Adelman & Rayle, 2000) promotes them as a logical choice when

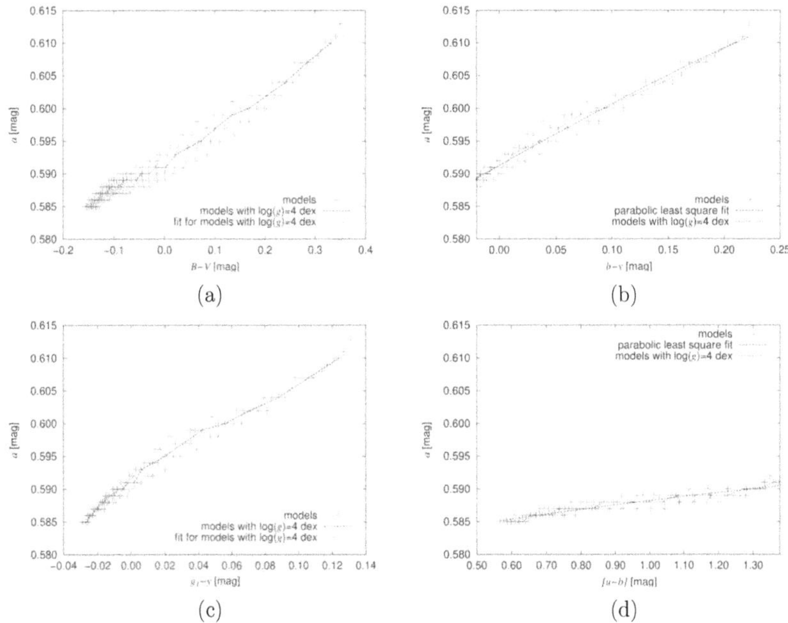

Figure 36: Temperature and gravity dependence along the main sequence for various observational indicators of T_{eff}. Colours were computed from ATLAS9 type model atmospheres with T_{eff}=[7000,15000] K for $\log g = [3.5,4.5]$ dex and solar metallicity. For the upper right panel, models were restricted to $(b-y) \geq -0.02$ while for the lower right panel, models were restricted to [u-b] ≤ 1.38. The zero point of the $(g_1 - y)$ colour was set to coincide with the zero point of $(b-y)$ at a $\log g$ of 4 dex (a constant of 0.269 had to be added to the output value of $(g_1 - y)$). Models with $\log g$ of 4 dex have been connected by straight lines to indicate the shape, zero-point, and slope of the normality line. For the cooler models (towards the right-hand side within each panel) the true ZAMS and thus the normality line a_0 are located slightly below this line (cf. Fig. 37). For the left-side panels a cubic least square fit through the $\log g = 4$ dex models is displayed, while right-side ones include a parabolic fit through all the models displayed.

testing a synthetic photometric system. Several grids of ATLAS9 model atmospheres were computed with the Stellar Model Grid Tool (SMGT, see Heiter et al., 2002b, for a description) using the line opacities and the ATLAS9 code of Kurucz (1993), unaltered except for the convection treatment (Smalley & Kupka, 1997; Heiter et al., 2002b). In fact, both Smalley & Kupka (1997) and Heiter et al. (2002b) recommend the use of a convection model in ATLAS9 which predicts inefficient convection for mid to late A-type stars ($T_{\rm eff} \leqslant 8500$ K). Therefore, the model of Canuto & Mazzitelli (1991) were used which allows a better reproduction of Strömgren colours of A-type stars than the original models of Kurucz (1993), as shown in Smalley & Kupka (1997). For models with $T_{\rm eff} > 8500$ K, where convection has only negligible influence on temperature gradients and colours, our models are virtually identical to those from the original grids published by Kurucz (1993). Anyway, a comparison with model atmospheres based on different convection models was done and the conclusion was drawn that for any of the convection treatments available for ATLAS9 (Castelli et al., 1997; Heiter et al., 2002b) the a values change only by up to +3 mmag for the coolest models in the grids and remain completely unaltered for models with $T_{\rm eff} > 8500$ K. Thus, no important bias is introduced into our calibration tests by selecting a particular convection model. The ATLAS9 grids have a spacing of $\Delta \log g = 0.5$ dex which is slightly too coarse for the purpose. Hence, also intermediate models were included in the computations by using a spacing of $\Delta \log g = 0.25$ dex. A $\log g$–range from 2.5 dex to 4.5 dex and a $T_{\rm eff}$–range from 7000 K to 15 000 K (with a spacing of 250 K) was studied. Model atmospheres assuming one of the following three metallicities were investigated: −0.5, 0, and +0.5 dex, where $[{\rm M/H}] = 0$ dex represents solar abundance and elements heavier than He are scaled by ±0.5 dex in the other cases. A constant value of 2 km s^{-1} as in the standard grid of Kurucz (1993) was used for the microturbulence.

To transform the theoretical colours into the frame of observed colours, Kurucz (1993) corrected the zero-point of his synthetic *uvby* system so as to match c_1, m_1, and $(b-y)$ of Vega. A similar procedure is necessary to compare calculated a values from the synthetic Δa system directly to observations. A value of 0.6 was added to the synthetic a value computed from convolving the effective transmission functions with the fluxes from ATLAS9 model atmospheres. The numerical values obtained are then close to (Maitzen & Vogt, 1983, see their Table 1 and Equation 1). One gets $a \sim 0.594$ mag for a main sequence star with $(b-y) = 0.000$ mag and solar metallicity from the synthetic photometric system. For the Δa system itself the specific value of a and thus any zero-point correction related to it are irrelevant, because the quantity of interest is the difference of a measured a

value to the normality line a_0 (Sect. 3). Hence, it is much more important to show that the synthetic Δa system recovers the observed dependencies of the Δa index from different indicators of effective temperature, surface gravity, and metallicity.

Figure 36 illustrates the temperature (and gravity) dependence of the a index as a function of various experimental indicators of $T_{\rm eff}$. For $(B-V)$ and (g_1-y) the entire temperature range is displayed while a cut-off was introduced for the other two cases through requiring that $(b-y) \geqslant -0.02$ (i.e. $T_{\rm eff} \lesssim 11\,000$ K) and [u-b] $\leqslant 1.38$ (i.e. $T_{\rm eff} \gtrsim 9250$ K). Models with $\log g = 4$ dex have been connected with straight lines to provide a proxy for the normality line a_0 of the synthetic photometric system. Note that the output colours have been rounded to 1 mmag accuracy as in Kurucz (1993). The actual run of the colours is continuous and smaller magnitude differences can hardly be assigned a real physical meaning within the current state of modelling.

For the comparison of the synthetic Δa system to observational ones, the results for Galactic field stars (Sect. 10) were used. The differences for the Δa values were found to be in the range of 2 to 3 mmag. The following relation for the normality line was found:

$$a_0 = G_0 + G_1(b-y) + G_2(b-y)^2 \tag{28}$$

with $G_0 = 0.594$, and where $0.086 < G_1 < 0.105$ as well as $-0.050 < G_2 < -0.150$ hold for $-0.120 < (b-y) < +0.200$ mag. The 1σ level around the normality line was found to be between 2.9 and 5.1 mmag. These values are a superimposition of the internal measurement errors and the (observed) natural bandwidth. On the other hand, the colours from model atmospheres ranging the main sequence band from $\log g$ of [3.5,4.5] dex for all models with a $T_{\rm eff}$ of [7000,15000] K and solar metallicity, and for which $(b-y) \geq -0.02$ mag, yield a $G_0 = 0.591$ (with less than 0.5 mmag error), while $G_1 = 0.0985 \pm 0.0056$ and $G_2 = -0.044 \pm 0.030$, when fitting a least square parabola through the model colours (see Fig. 36). Hence, the $(b-y)$ dependence of the experimental Δa systems is reproduced very well.

For the $[u-b]$ relation, Maitzen (1985) lists $G_1 = 0.024$ for 22 bright unreddened stars with a 1σ level of 4.5 mmag. For this correlation, the results in Fig. 36 imply a more flat dependence of $G_1 = 0.0098 \pm 0.0018$ (and a G_2 of -0.0022 ± 0.00097). However, the rather small slope is very sensitive to the precise definition of the sample: including giants with $\log g \geqslant 2.5$ dex would raise G_1 to 0.0161, whereas reducing the range of $[u-b]$ from an upper limit of 1.38 to 1.20 while keeping only the main sequence band models with $\log g \geqslant 3.5$ dex, as in the first case, would increase it to

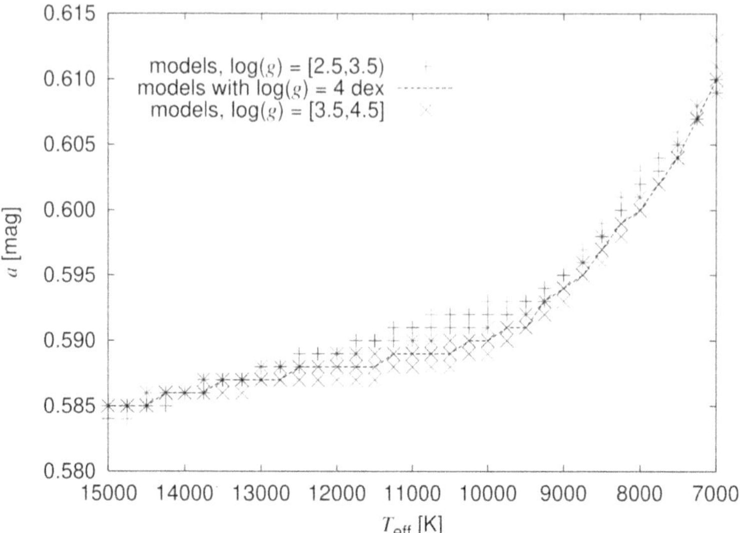

Figure 37: Luminosity dependence of the a index. Colours from ATLAS9 type model atmospheres with T_{eff}=[7000,15000] K for $\log g$ = [2.5,4.5] dex and solar metallicity. Models above the main sequence band and thus at lower surface gravity are indicated by different symbols. Asterisks indicate the overlap of both ranges. Output accuracy is limited to 1 mmag. Models with $\log g$ = 4 dex have been connected by straight lines to indicate the normality line a_0.

$G_1 = 0.0211 \pm 0.0028$. Thus, within the overall uncertainties expected for such kind of a weak dependence on $[u - b]$ the latter is reproduced sufficiently well.

Figure 36 also shows the correlation of a with the temperature indicators $(B - V)$ and $(g_1 - y)$. Their dependency can easily be studied as before, but is unlikely to reveal more information beyond the uncertainties introduced by the colour transformation required to compare the different filter systems used in observations and the synthetic systems of Kurucz (1993).

14.1.3 Luminosity effects

Because each of the colour relations presented in Fig. 36 is also affected by surface gravity within the range of effective temperatures populated by the CP stars, the direct dependence of a on T_{eff} was investigated as well (see

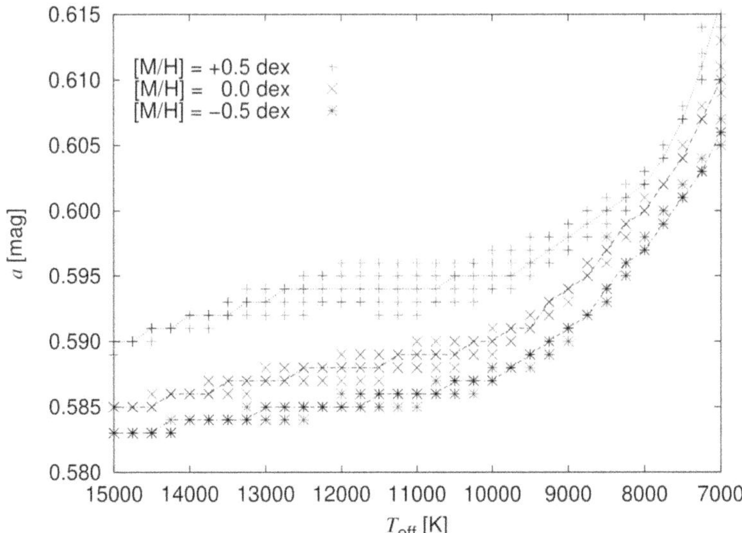

Figure 38: Metallicity dependence of the a index. Colours from ATLAS9 type model atmospheres with $T_{\rm eff}$=[7000,15000] K for $\log g$=[3.5,4.5] dex. Metallicities are scaled with respect to the solar abundances used in Kurucz (1993). Again, models with $\log g = 4$ dex have been connected by straight lines to indicate the normality line a_0.

Fig. 37). Clearly, models with lower surface gravity have higher a values. The effect of surface gravity on the a index is largest for the late B-type stars with effective temperatures around $T_{\rm eff} \sim 10\,500$ K. The width of the band of standard stars as induced by surface gravity for a given metallicity is between 2 and 5 mmag. This confirms the results of Sect. 14.1.2 and is in agreement with the observational data quoted therein.

One hast to note here that the step size in $\log g$ for model sequences shown in both Figs. 37 and 38 is 0.25. However, as the output of the photometric indices has been truncated to 1 mmag, it turns out that the Δa dependence on $\log g$ is too weak to show up more prominently. Hence, many models overlap in the Figures due the assumed output accuracy. The sensitivity of Δa to $\log g$ slightly depends on $T_{\rm eff}$ and metallicity, as the number of apparent points in Figs. 37 and 38 reveals as well.

14.1.4 Metallicity effects

Metallicity has an effect on the a index which is actually more important than that of luminosity (surface gravity). Figure 38 compares the main sequence band for solar metallicity with models having over- and underabundances of ± 0.5 dex for all elements heavier than He. An underabundance of -0.5 dex as in the Magellanic Clouds (Dirsch et al., 2000) yields a shift of the normality line of -3 mmag. The size of this shift is quite constant over the entire $T_{\rm eff}$ range relevant for CP stars and also within the entire luminosity range expected for the main sequence band. On the other hand, an overabundance of $+0.5$ dex yields a larger shift of between $+3$ and $+6$ mmag with a maximum for the late B-type stars. This behaviour gives already some hint on the nature of the flux depression at 5200Å in agreement with Adelman & Rayle (2000) who found that line opacities of ATLAS9 models with metal overabundances of $+0.5$ and $+1.0$ dex predict some extra line blanketing in this region. In turn, due to the good agreement of ATLAS9 model fluxes in this wavelength region with observations for mildly peculiar stars, which have underabundances or overabundances of up to about 0.5 dex (Adelman & Rayle, 2000), and due to the satisfactory agreement of the derived synthetic Δa system with systems used in observations, an important conclusion can be drawn: application of Δa photometry to the Magellanic Clouds (Sect. 19.3) will lead only to a small bias, with a size of about -3 mmag relative to the observations made for the solar neighbourhood, and the same will hold for more remote targets that have a similar metallicity range.

14.2 The cool CP and λ Bootis stars

After establishing the synthetic Δa photometric system (Sect. 14.1.1), model atmospheres were computed with individual abundances for a representative sample of CP stars. This will either confirm or redetermine their input parameters through comparisons with photometric, spectrophotometric, and high resolution spectroscopic data. The final models obtained from this procedure were used to compute synthetic Δa indices which were compared with observations.

The determination of abundances for CP stars is not straightforward. Their magnetic fields together with inhomogeneous surface abundance distributions makes a reliable analysis very difficult. One of the main conclusions drawn from previous works on this field has been the necessity to build specific model stellar atmospheres for CP stars using state-of-the-art opacity data. The increase in available computer power and advances in computational algorithms during the last two decades have now brought this problem into the realm of workstations and personal computers. Moreover, stellar atmosphere modelling can take advantage of data bases for atomic line transitions devoted to stellar atmosphere applications such as Kurucz (1992) and the VALD project (Kupka et al., 1999). Among the current projects for the computation of model atmospheres several are based on an opacity sampling approach. The PHOENIX code (Hauschildt et al., 1999) has been developed for computations of model atmospheres throughout the entire HRD. However, for the computation of small grids of model atmospheres with individual abundances the opacity distribution function (ODF) approach remains preferable due to its higher computational efficiency. The main limitation of this approach is that it is unsuitable for taking vertical stratification into account. However, as it was decided to consider models with a chemically homogeneous composition first to explore capabilities and limitations, and to provide a starting point for more elaborate analyses later on, the approach of Piskunov & Kupka (2001) was taken to construct model atmospheres with individual abundances for CP stars.

The observed behaviour of Δa is reproduced for several types of CP stars: models for Am stars show negligible (or marginally positive) values of a few mmag, while for λ Bootis stars – and for metal deficient A-type stars in general – negative values (as low as -12 mmag in one case) were obtained. For the coolest CP2 stars with effective temperatures below about 8500 K, mild ($\sim +10$ mmag) to moderately large ($\sim +30$ mmag) flux depressions in agreement with observations were obtained. However, Δa values for slightly hotter members of the CP2 group (for which still $T_{\rm eff} <$ 10 000 K) are underestimated from these new models. The effect of the microturbulence parameter on the Δa index is revisited and its different role in various types of CP stars for reproducing the flux depression at 5200Å is explained.

14.2.1 Model atmospheres for CP stars

The model atmospheres applied in this study have been computed with a variant of the ATLAS9 model atmosphere code which uses line and Rosseland mean opacities calculated for individual chemical composition, as described in Piskunov & Kupka (2001). The models assume plane parallel geometry, hydrostatic equilibrium, local thermal equilibrium (LTE), horizontal homogeneity, and a homogeneous chemical composition. An opacity distribution function (ODF) approach (Kurucz, 1970) is used for the computation of line opacities for these models. Piskunov & Kupka (2001) have also confirmed the equivalence of their model atmospheres to those of Kurucz (1993), if identical element abundances are used. Further model atmospheres have been computed using the standard ATLAS9 code (except for the treatment of convection, see below) and the line opacities of Kurucz (1993) with a solar (Anders & Grevesse, 1989) or scaled solar abundance. These were used to initiate our computational procedure and to provide a reference for the effect of including an individual chemical composition in the modelling of the atmosphere of each target star.

The new model atmospheres with individual chemical composition differ from those used in earlier studies such as Maitzen & Muthsam (1980) by enhanced numerical resolution and the usage of much more accurate and complete atomic line data provided such as Kurucz (1992) and the VALD project (Kupka et al., 1999). Recent enhancements to line opacities include the contributions of rare earth elements from neutral, singly, and doubly ionised stages, of strong and intermediately strong lines from iron peak elements, and the addition of millions of weak lines from predicted energy levels of iron peak elements (Kurucz, 1992). The new line data allows to take full advantage of the higher resolution and the better signal to noise ratio of modern spectroscopic observations. Contrary to earlier studies reduced overabundances have been found when using the current atomic data for both the analysis of individual spectral lines and the construction of model atmospheres. For example, compare the abundance analysis of γ Equ (HR 8097) by Adelman (1973a) and Ryabchikova et al. (1997), who also give examples of how including the effects of hyperfine and Zeeman splitting naturally yields even lower overabundances. The new line data provide sufficiently accurate description of line blanketing to allow a much better though not yet fully satisfactory match of the spectral flux distribution from the far UV to the infrared. This has been confirmed by detailed comparisons with flux observations of Vega (Castelli & Kurucz, 1994; Piskunov & Kupka, 2001). It has also been found to hold at least as long as chemical peculiarities are moderate enough to be accounted for by simply scaling the solar metallicity (Adelman & Rayle, 2000) and convection is not strong enough to form a granulation pattern as in the solar case. See also the discussion on the effects of convection treatments in Smalley & Kupka (1997).

However, despite the success of the revised atomic data in explaining the spectrophotometry of stars with a chemical composition similar to the Sun apart from a scaling factor, even an overall fit of the flux distribution using model atmospheres with scaled solar chemical composition is generally not sufficient to match flux depressions observed in CP stars such as the 5200Å feature (Adelman & Rayle, 2000). Moreover, extreme peculiarities may even change the atmospheric structure (Piskunov & Kupka, 2001). For that reason it was decided to compute models with individual chemical compositions for our study (Table 17) following the computational procedure described in detail in Piskunov & Kupka (2001). In general, for larger deviations a reanalysis of the element abundances may have to be considered to obtain a model which predicts fluxes and rectified spectra that match the observational data. In the case of HR 7575, for which a larger change in $\log g$ was favoured, it was verified that the derived element abundances from the final model are still within the error range expected for the original model.

In some cases, abundances and input $T_{\rm eff}$ and $\log g$ had to be changed, although mostly within the limits mentioned above. The model construction was iterated this way until a more satisfactory match with the observations had been achieved. The best matching model atmospheres obtained from this procedure were used to compute synthetic Δa indices and the latter were compared with the observed Δa data. In a final stage, detailed synthetic spectra for the region measured by the Δa system were computed and compared with observations to probe the reliability of the chemical composition and fundamental parameters used as well as the obtained model atmosphere structure, and in particular to detect the main contributing species to the enhanced line blanketing in the region of the 5200Å flux depression.

For this procedure the use of Balmer line profiles was avoided in order to constrain the model parameters due to the uncertainties in the broadening mechanisms and the effects from the convection treatment found for stars with $T_{\rm eff} < 9000$ K (Smalley et al., 2002). Further uncertainties are introduced by the complex dependence of the Balmer lines on both $T_{\rm eff}$ and $\log g$ in that region (Heiter et al., 2002b), and particularly by their ambiguity near the Balmer maximum which may lead to severe errors when deriving fundamental parameters (Smalley et al., 2002).

One has to point out at this stage effects of a magnetic field on the atmospheric structure, with the exception of the enhanced line blanketing created by the Zeeman splitting of lines were considered. The latter has been modelled by a pseudo-microturbulence in stars with known large magnetic fields ($>$ 2 kG). This concept dates back to Adelman (1973b) and Hensberge & De Loore (1974). Due to indications of strong stratification in the coolest stars of the sample (Ryabchikova et al., 2005) it was assumed that convection has a negligible influence on the temperature gradient in all our sample stars. Hence, similar to Piskunov & Kupka (2001), for most calculations the convection model of Canuto & Mazzitelli (1991, CM) was used, because it predicts the most inefficient convection among

Table 16: Cool CP stars for this study (upper panel). The spectral type classification has been taken from Renson & Manfroid (2009) and Gray & Corbally (1993). Values for T_{eff} and $\log g$ are those redetermined in Section 14.3. Data for the surface magnetic field and microturbulent velocities ξ_{t} are taken from the sources given in the last column: (1) Ryabchikova (2017, priv. comm.), (2) Ryabchikova et al. (1999a), (3) Kato & Sadakane (1999), (4) Caliskan & Adelman (1995). For β CrB, 33 Lib, and HR 7575, ξ_{t} has been enhanced to account for magnetic intensification effects (cf. Sect. 14.2.1 and Ryabchikova et al., 1999a). Additional target stars, of λ Bootis type, are listed in the lower panel. In their case, model data and fluxes have been taken from Heiter et al. (1998).

HD	Star name	Spectral type	Peculiarity	T_{eff} [K]	$\log g$ [dex]	ξ_{t} [km s^{-1}]	B_{s} [kG]	Ref
108642	HR 4750	hA2mF0	CP1	8200	4.0	4.0	~ 0	1
108651	17 Com B	hA2mF1	CP1	7900	4.3	4.0	~ 0	1
137909	β CrB	A9 SrEuCr	CP2	8000	4.3	4.0	5.7	2
137949	33 Lib	F0 SrEu	CP2 (roAp)	7550	4.3	4.0	4.9	2
188041	HR 7575	A6 SrCr	CP2	8700	4.5	1.75	3.6	2,3
196502	73 Dra	A2 SrCr	CP2	8900	4.0	2.0	2.0	2
201601	γ Equ	A9 SrEu	CP2 (roAp)	7700	4.2	2.0	4.0	2
204411	HR 8216	A6 Cr	CP2	8400	3.5	2.0	0.5	4
183324	35 Aql	A0 Vb	λ Boo	9300	4.3	3.0	~ 0	
192640	29 Cyg	A0.5 Va$^-$	λ Boo	7800	4.0	3.0	~ 0	

the various models available for use with ATLAS9 (Heiter et al., 2002b). As a consequence, the temperature gradients obtained for the investigated CP stars are either nearly or even completely radiative. This and several other limitations of the modelling approach and its ability in explaining the observational data are discussed in more detail in Section 14.6.

14.2.2 Selected sample of stars

To find out whether model atmospheres based on individual element abundances are capable to reproduce the 5200Å feature, and in particular the Δa index, well studied members from the major groups of CP stars were selected to perform case studies. The sample of stars was supposed to fulfill the following requirements: a) multiband photometry for at least the Johnson UBV and Strömgren $uvby\beta$ – and preferably also for the Geneva 7-colour system – had to be available, as well as b) spectrophotometric measurements ranging from the near UV to the red region. Moreover, c) an abundance analysis and determination of fundamental parameters, preferably based on model atmospheres and atomic data similar to Kurucz

Table 17: Individual abundances (in dex, relative to total number of atoms for the Sun) for all CP1 and CP2 stars discussed; the solar values are from Anders & Grevesse (1989). Last digits for Dy, Er, Yb, and Lu have no significance. Only Yb and Lu for HD 137909, Yb for HD 137949, as well as Dy and Er for HD 201601 are based on actual abundance determinations of these heaviest elements. The Ne abundance has not been determined for any of the target stars and thus exhibits the scaling factor used for other undetermined element abundances. The solar He abundance was assumed in all model atmosphere calculations.

Elem.	108642	108651	137909	137949	188041	196502	201601	204411	Sun
He	+0.00	+0.00	+0.00	+0.00	+0.00	+0.00	+0.00	+0.00	−1.05
C	−0.39	−0.23	−0.72	−0.70	−0.50	−0.50	−0.52	−0.70	−3.48
N	+0.50	+0.50	−0.93	−0.60	−0.50	−0.50	−0.60	−0.60	−3.99
O	−0.63	−0.43	−0.89	−0.39	−0.50	−0.50	−0.79	−1.30	−3.11
Ne	+0.50	+0.50	+0.50	+0.50	+0.50	+1.00	+0.50	+0.50	−3.95
Na	+0.38	+0.24	+0.28	−0.19	+0.50	+1.00	+0.44	+0.50	−5.71
Mg	−0.28	−0.31	+0.50	+0.50	+0.01	+1.00	+0.17	−0.38	−4.46
Al	+0.50	+0.50	+0.50	+0.50	+0.50	+1.00	+0.64	+0.50	−5.57
Si	+0.21	+0.43	−0.31	−0.01	+0.50	+0.01	+0.07	−0.05	−4.49
S	+0.50	+0.50	+0.50	+0.50	+0.50	+1.00	+0.09	+0.50	−4.83
Ca	−0.47	−0.25	+0.18	+0.08	+0.78	−0.50	+0.21	+0.44	−5.68
Sc	−0.34	−0.27	+0.50	+0.50	−0.02	−1.25	−0.51	−0.70	−8.94
Ti	+0.01	+0.09	+0.50	+0.50	+0.77	+0.33	+0.06	+0.51	−7.05
V	+0.52	+0.57	+0.50	+0.94	+0.50	−0.04	+0.96	−0.30	−8.04
Cr	+0.23	+0.40	+1.77	+1.62	+2.31	+2.67	+0.89	+1.33	−6.37
Mn	+0.63	+0.65	+0.50	+0.50	+2.07	+0.37	+1.02	+0.61	−6.65
Fe	+0.07	+0.08	+0.27	+0.02	+0.86	+1.07	+0.07	+0.33	−4.37
Co	+0.68	+0.59	+0.50	+0.50	+0.50	+0.44	+1.11	+0.58	−7.12
Ni	+0.67	+0.72	+0.50	−0.01	+0.38	+0.58	−0.14	+0.05	−5.79
Cu	+0.93	+0.73	+0.50	+0.50	+0.50	+1.00	+0.50	+0.50	−7.83
Zn	+0.42	+0.35	+0.50	+0.50	+0.50	+1.00	+0.50	+0.50	−7.44
Sr	+0.83	+0.55	+0.50	+0.50	+0.51	+1.00	+2.18	+0.60	−9.14
Y	+0.94	+0.70	+0.50	+0.50	+0.50	+0.04	+1.43	−0.14	−9.80
Zr	+0.88	+0.32	+0.50	+0.50	+0.50	+0.75	+0.76	+0.40	−9.44
Nb	+0.50	+0.50	+0.52	+0.52	+0.50	+1.00	+1.77	+0.52	−10.62
Mo	+0.50	+0.50	+0.50	+0.50	+0.50	+1.00	+1.87	+0.50	−10.12
Ba	+0.60	+0.80	+0.51	+0.31	+0.50	+1.00	+0.85	+0.97	−9.91
La	+1.05	+1.01	+2.32	+2.32	+2.76	+0.67	+1.23	−0.28	−10.82
Ce	+1.67	+1.93	+2.99	+2.49	+2.73	+0.77	+1.29	+0.50	−10.49
Pr	+0.50	+0.50	+2.03	+2.13	+2.28	+1.48	+1.35	+1.79	−11.33
Nd	+0.89	+0.97	+2.09	+2.19	+0.50	+1.00	+1.37	+1.00	−10.54
Sm	+0.85	+0.82	+2.44	+2.94	+0.50	+1.00	+1.51	+0.74	−11.04
Eu	+1.38	+1.39	+3.03	+2.53	+4.63	+2.73	+1.33	+0.33	−11.53
Gd	+1.25	+1.27	+0.52	+0.52	+3.75	+1.44	+1.57	+0.52	−10.92
Dy	+0.54	+0.54	+0.54	+0.54	+0.54	+1.00	+1.80	+0.54	−10.94
Er	+0.51	+0.51	+0.51	+0.51	+0.51	+1.01	+1.11	+0.51	−11.11
Yb	+0.56	+0.56	+2.46	+2.16	+0.56	+1.00	+0.56	+0.56	−10.96
Lu	+0.58	+0.58	+1.28	+0.58	+0.58	+1.08	+0.58	+0.58	−11.28

(1993), and d) most importantly, measurements of the Δa index had to be available. For about half of the target stars also new spectra were taken to analyse the contributing species in and around the 5200Å region. In a few cases it was possible to supplement these data by spectrophotometric scans of the region such as have frequently been used in earlier studies of the 5200Å feature (Maitzen & Muthsam, 1980). Section 14.5.4 includes two examples to judge the amount of line-blanketing accounted for by the new model atmospheres. To minimise the influence of reddening we have chosen only bright, close-by target stars. Hence, $E(B-V)$ is most likely less than 0.01 mag for all objects we have studied here.

In Table 16 the basic parameters of all target stars for this study are listed which have a T_{eff} of less than 10 000 K. They comprise two CP1 stars, four very cool CP2 stars (mid A- to late A/early F- type), and two slightly hotter CP2 stars with a very strong 5200Å feature.

In addition, two λ Bootis stars (35 Aql and 29 Cyg) are listed. For 35 Aql and 29 Cyg these underabundances amount to -1.5 dex and -2.0 dex for the iron peak elements with respect to solar values. Thus, they represent the opposite case to the CP1 and CP2 stars. For both stars neither further investigations of their basic parameters nor model computations were necessary, because they had already been done by Heiter et al. (1998) with exactly the same software tools used here. Hence, it was possible to directly use the output fluxes calculated from their model atmospheres for this study of the Δa index. Refer to Heiter et al. (1998) for further details on parameter determinations of these stars, in particular for metallicities and model calculations.

14.3 Individual models: parameter determination with observed fluxes

In the following, the consistency checks and improvements of the parameter determinations for our CP1 and CP2 target stars based on the new model atmospheres with individual abundances are described. The complete list of element abundances as used for the final models is given in Table 17. Synthetic as well as observed multiband photometry of the targets have been collected in a separate table available on request from the author, as the main focus has been the investigation of the capability of the models to reproduce observed spectrophotometry. The observational data for the comparison with model atmosphere fluxes was taken from the literature. If more than one measurement was found, the values were averaged. The errors of the means are in general smaller than those of individual

Table 18: Comparison of the spectroscopic results of Savanov (1996) for T_{eff} (in K) and $\log g$ (in dex) of our target CP1 stars with photometry. The $uvby\beta$ photometry from Paunzen (2015) and the calibrations by Napiwotzki et al. (1993) and Smalley (1993) were used for the photometric parameter determinations. The microturbulent velocity ξ_t is given in km s^{-1} as found by Savanov (1996) from Fe II lines.

HD	$uvby\beta$			Savanov (1996)		
	T_{eff}	$\log g$	[M/H]	T_{eff}	$\log g$	ξ_t
108642	8087	4.01	0.25	8200	4.0	5.3
108651	8068	4.20	0.36	8000	4.3	5.6

measurements, although for some stars there is considerable variation over their rotational period (see also discussions in Sects. 14.4 and 14.5.3). It has to be emphasized that almost all spectrophotometric observations are normalised to the flux at 5000Å which is rather close to the investigated 5200Å depression. This might introduce an error source when comparing those data to synthetic spectra. Therefore also a comparison of these data with a normalisation at 5556 Å was performed and no significant differences were found, as it is discussed in Section 14.3.1.

14.3.1 CP1 stars

Detailed model atmospheres for two Am stars were computed: HR 4750 (HD 108642) and 17 Com B (HR 4751, HD 108651). Both are members of the Coma cluster (Guerrero et al., 2015). HR 4750 is an SB 2 type binary (Boesgaard, 1987) with a period of 11.78 days while 17 Com B is of SB 1 type with a period of 68.29 days (Savanov, 1996) and is occasionally misclassified as an Ap star.

Table 18 lists the fundamental parameters of both stars obtained from photometry and from the spectroscopic analysis of Savanov (1996). Microturbulent velocities of 5 to 6 km s^{-1} were found for a number of atomic species. For both objects, the starting point was a solar composition as in Kurucz (1993) and Anders & Grevesse (1989), for which elements heavier than He had in addition been scaled by +0.5 dex. This composition was modified by using the element abundances obtained from the spectroscopic analysis published in Table 6 of Savanov (1996) wherever possible, with the exception of (the less reliable result for) Al. For O an additional, more recent result based on the same observations was used. In most cases where

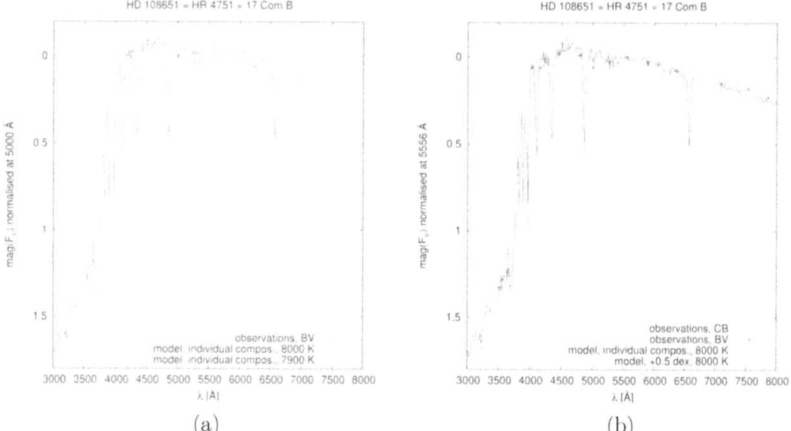

Figure 39: Spectrophotometric observations of 17 Com B are compared to fluxes from model atmospheres with individual chemical compositions for two values of T_{eff}: 7900 K and 8000 K. Note the effect of different normalisations: the standard one at 5000Å (left panel) is compared to an alternative one at 5556Å (right panel).

both neutral and singly ionised species of an element had been analysed, their average was used, except for Si and Cr, where results from singly ionised species were preferred (changes would have been negligibly small for the applications, if averages were considered for them instead). A value of 4 km s^{-1} for the microturbulence parameter for any of our model atmospheres for HR 4750 and 17 Com B (Table 16) was taken, as it provides a safe lower estimate for the different atomic species and is readily available also from the model grids published by Kurucz (1993).

Because the agreement between photometric and spectroscopic values of T_{eff} and $\log g$ is quite good (Table 18), the latter ones were taken to compute models as well as synthetic photometric colours with the chemical composition as just described. The synthetic photometric colours for all the models for HR 4750 and 17 Com B turned out to be in agreement with observations to within the uncertainties of colour transformations for *individual stars*.

Figure 39 compares the spectrophotometric observations of Böhm-Vitense & Johnson (1977) for 17 Com B with fluxes from model atmospheres with individual chemical compositions for two values of T_{eff}: 7900 K and 8000 K. The upper panel of the figure assumes that fluxes are normalised at 5000Å such that mag(F_ν) equals zero. This choice represents the standard convention in visual spectrophotometry. With this normalisation, a slightly lower T_{eff} of 7800 K to 7900 K appears preferable to the value of 8000 K

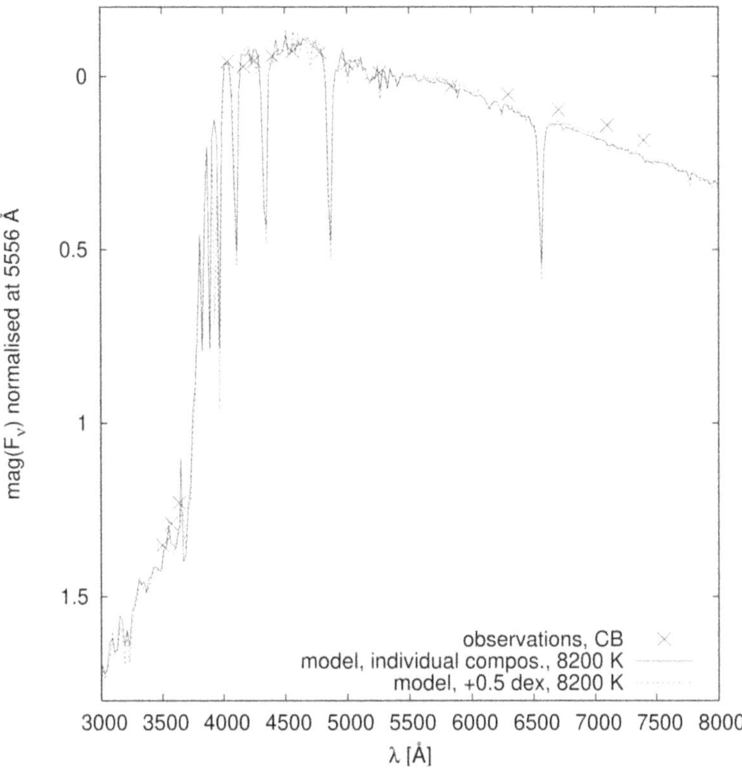

Figure 40: Spectrophotometric observations of HR 4750 from Clampitt & Burstein (1997) are compared to fluxes from model atmospheres with an individual chemical composition and with a composition scaled with respect to the solar one.

used in the abundance analysis of Savanov (1996). However, as the normalisation to an alternative wavelength of 5556Å reveals (right panel), a lowering of $T_{\rm eff}$ by 100 K or even more might actually turn out to be just an overcompensation of the normalisation procedure with its finite accuracy. This claim is corroborated by a comparison to observations by Clampitt & Burstein (1997) in the same panel: note that the apparently beneficial effect of normalising the Böhm-Vitense & Johnson (1977) data at the wavelength of 5556Å is questionable, because a shift in the opposite direction is needed to improve the match with the data of Clampitt & Burstein (1997) which had originally been normalised to 5556Å. Consequently, because of the limitations of the present spectrophotometric data, modifications of the normalisation wavelength are not likely to provide any improvements of the analysis and it was decided to use the traditional wavelength of 5000Å as the normalisation point for all subsequent work.

Figure 40 displays a comparison of spectrophotometric observations of HR 4750 from Clampitt & Burstein (1997) with the model fluxes. Note that the luminosity of the late F- or early G-type companion of HR 4750 is much smaller than that of the Am type primary of the system. For fundamental parameters chosen as in Table 16, i.e. with $T_{\rm eff}$ and $\log g$ from Savanov (1996) as well as a $\xi_{\rm t}$ of 4 km s^{-1}, the match appears slightly better than for 17 Com B (right panel of Fig. 39). Note that the fluxes from a model based on individual abundances and one based on a solar composition scaled by +0.5 dex can hardly be distinguished on the level of current spectrophotometric precision. The same can be concluded for the case of 17 Com B. Hence, it was decided to use the parameters in Table 16 and the abundances described in this Subsection in the studies of the 5200Å depression.

14.3.2 CP2 stars with $T_{\rm eff} \leqslant 8500$ K

The SrEu subclass of CP2 stars includes the magnetic CP stars which have some of the lowest $T_{\rm eff}$ among all CP stars. Two of them were chosen for this study: 33 Lib (HD 137949) and γ Equ (HD 201601). Both also belong to the rapidly oscillating Ap (roAp) stars (Bigot & Kurtz, 2011) which are characterised by oscillations with amplitudes of about a few mmag and periods of less than 25 minutes. To complement the sample two further objects were included which are quite distinct from each other. β CrB (HD 137909) shares many similarities with roAp stars, in particular its range of colour indices and low effective temperature while HR 8216 (HD 204411) is distinguished from these three by its higher $T_{\rm eff}$ and the much lower overabundance of Eu.

For the initial model atmosphere of β CrB, $T_{\rm eff}$, $\log g$, (pseudo-) microturbulence, and the chemical composition were taken from Ryabchikova et al. (1999a) supplemented by additional element abundances determined by Ryabchikova (2017, priv. comm.; see also Table 17). Scaled solar abundances as in the model of Ryabchikova et al. (1999a) have been assumed for all the other elements. After comparing fluxes from model atmospheres with individual composition to the observations of Adelman et al. (1989), it was found that a slightly increased $T_{\rm eff}$ allows a better match with the data. The upper left panel of Fig. 41 shows the fluxes of the observations as well as those of the best fitting models obtained with either composition. Their parameters are summarised in Table 19. Clearly, the model with individual composition achieves a much better match for the entire range from the Balmer jump to about 6500Å than its counterpart with a scaled solar metallicity. Changing T_{eff} for the scaled composition cannot emulate this effect, because the predictions of the UV fluxes get worse. On the other hand, a T_{eff} of 7750 K for the model atmosphere with individual composition yields worse results than our choice of 8000 K in both the UV and blue region as well as in the red region and most of the near IR. A slightly higher T_{eff} of 7850 K would match the far red/IR region best, but would not be as satisfactory for the remaining part of the visual region, nor the UV, compared to the hotter model shown in Figure 41. Hence, the latter represents the best model which can be found under the assumption of chemical homogeneity to within a fitting uncertainty of ± 100 K. Overall, the new model is a considerable improvement over its predecessor: if abundances are assumed to be merely scaled solar ones, it is difficult to match more than one or two spectrophotometric regions of β CrB with a size of a few 100Å simultaneously. The new model matches a region of 3000Å ranging from the UV to the red quite satisfactorily. None among the chemically homogeneous models is able to match the slope of the near IR region from 6800Å to 7800Å as well as the visual region at the same time.

For 33 Lib, abundances and initial model data were taken from Ryabchikova et al. (1999a) with additional element abundances from Ryabchikova (2017, priv. comm.; see Table 17 for a summary). The latter are based on a revision of the results of Gelbmann (1998). Scaled solar abundances as in Ryabchikova et al. (1999a) have been assumed for all other chemical elements. Similar to the case of β CrB, a slightly higher T_{eff} preferable for the model was found with individual chemical composition, and, in addition, also a slightly lower value for the surface gravity (see Table 19). A comparison of the fluxes of this new model to the observations of Adelman et al. (1989) and to the previous model, which was based on a scaled solar metallicity, is shown in Figure 41. The new model matches the observations remarkably well over most of the entire wavelength range from near UV to the near IR. Note here that Ryabchikova et al. (1999a) preferred a slightly different normalisation of the calculated fluxes of their 33 Lib model atmosphere, which is displayed in their Figure 4. While $\text{mag}(F_\nu) = 0$ at 5000 Å was enforced, they have determined the zero point so as to achieve a good match (in the least square sense) of the neighbouring data points as well. Indeed, the uncertainties of the observed fluxes prohibit a more accurate determination of the exact normalisation magnitude at 5000Å and

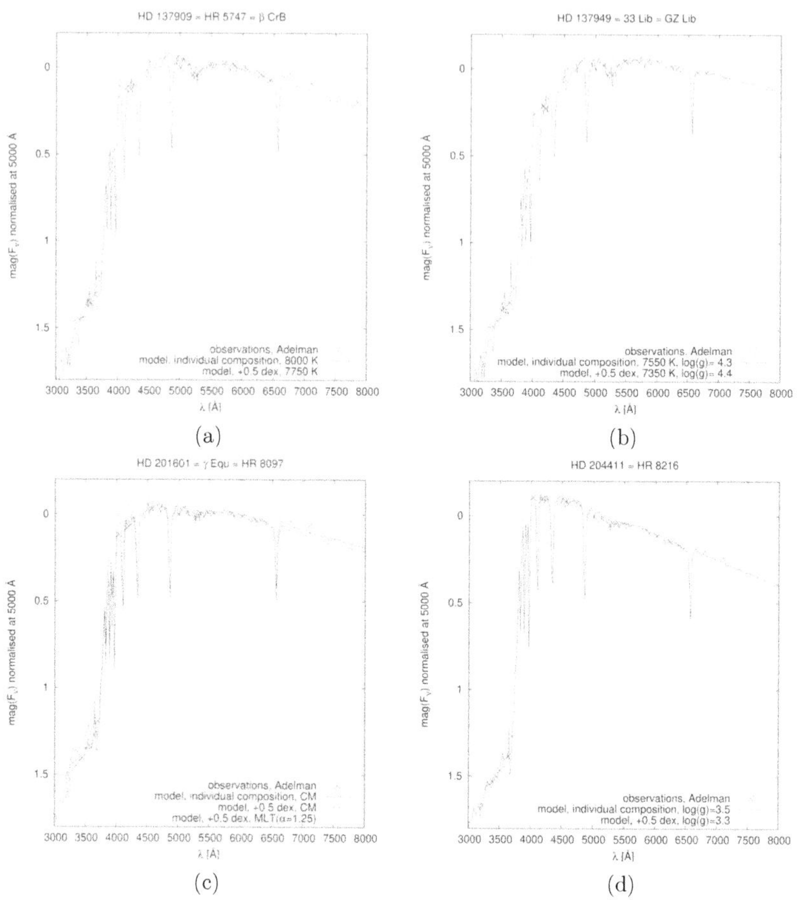

Figure 41: Upper left panel: spectrophotometric observations of β CrB are compared to fluxes from a model atmosphere with individual chemical composition and a model with enhanced metallicity (scaled solar abundances). Upper right panel: the same for 33 Lib. Lower left panel: the same comparison for γ Equ, but with an additional graph for a model atmosphere with scaled solar abundance, but a different treatment of convection. Lower right panel: a similar comparison as for β CrB done for HR 8216. Modelling and comparison details are discussed in the text.

Table 19: Parameters for model atmospheres of CP2 stars with a scaled solar composition (Anders & Grevesse, 1989), which has been used in spectroscopic analyses in the literature, are compared to those derived for the models with individual chemical composition. $T_{\rm eff}$ is given in K, other parameters are in dex. Literature parameters are from Ryabchikova et al. (1999a), except for HD 204411 (Caliskan & Adelman, 1995), and for the second rows for HD 188041 and HD 196502, which contain the results from Lyubimkov (1986) and Kato & Sadakane (1999), respectively. For the microturbulence $\xi_{\rm t}$ see Table 16. Details on individual abundances are found in the text and in Table 17.

HD	Literature parameters			This analysis		
	$T_{\rm eff}$	$\log g$	[M/H]	$T_{\rm eff}$	$\log g$	[M/H]
137909	7750	4.3	+0.5	8000	4.3	ind.
137949	7350	4.4	+0.5	7550	4.3	ind.
188041	8500	4.5	+0.5	8700	4.5	ind.
	8500	3.5	+1.0			
196502	8700	3.8	+1.0	8900	4.0	ind.
	8150	3.6	+0.0			
201601	7700	4.2	+0.5	7700	4.2	ind.
204411	8400	3.3	+0.5	8400	3.5	ind.

leave some freedom for the determination of the zero point of the calculations (Sect. 14.3.1 and Fig. 39). If the normalisation of Ryabchikova et al. (1999a) is used instead, a slightly lower value for $T_{\rm eff}$ of about 100 K would have been preferable for the new model with individual chemical composition. In any case, the new model predicts the various flux features of 33 Lib in the visual region considerably better than the model with scaled solar abundance. These conclusions hold independently of the particular convection model used for the model computations: even for an atmosphere based on a mixing length treatment of convection with a large value of mixing length to pressure scale height ($\alpha = 1.25$), the chosen $T_{\rm eff}$, $\log g$, and the Δa obtained would be the same within the uncertainties of the spectrophotometric data. Because 33 Lib is the coolest star in the entire sample, this gives some comfort concerning the independence of the results from the particular treatment of convection in the model atmospheres. This result was indeed expected from the lower convective efficiency found for model atmospheres with enhanced metallicity (Piskunov & Kupka, 2001).

For γ Equ the fundamental parameters and element abundances derived by Ryabchikova et al. (1997) were taken using solar abundances scaled by +0.5 dex for all other elements, except for C and O, for which redetermined values by Ryabchikova (2017, priv. comm.) have been used, and N, for which we have taken an estimate based on Roby & Lambert (1990). A summary of the final abundances is given in Table 17. Even without altering $T_{\rm eff}$ or $\log g$ the match of the observed fluxes of Adelman et al. (1989) to the fluxes from the new model has improved over its predecessor with scaled solar abundances, as can be seen from Figure 41. This result is once more essentially independent of the convection model used for our computations. To illustrate this remark the fluxes from model atmospheres with scaled solar composition based on both the mixing length theory (no overshooting, $\alpha = 1.25$) and the convection model of Canuto & Mazzitelli (1991) were included in Figure 41. Whereas a preference for one or the other convection model (for the $T_{\rm eff}$ and $\log g$ chosen here) cannot be supported from this comparison, the improvements from using an individual chemical composition in the calculations are much more obvious. Only in the IR the emergent (normalised) flux is still slightly underestimated by the new model, though less than by models with scaled solar composition.

The fundamental parameters and the element abundances for the model atmosphere for HR 8216 were taken from Caliskan & Adelman (1995). For species not included in their Table 2, solar abundances scaled by +0.5 dex were assumed, as used in their analysis, except for CNO, for which the estimates for CP2 stars derived by Roby & Lambert (1990) were used. A summary is shown in Table 17. As can be seen from Fig. 41, already the fluxes of the original model of Caliskan & Adelman (1995) match the observations of Adelman et al. (1989) very well, except for the underestimation of the flux in the far red and near IR. However, due to the milder overabundance of HR 8216 the new model hardly yields any improvements in comparison with models with a scaled solar abundance. One might argue whether a slightly larger surface gravity with a $\log g$ of 3.5 dex may be taken instead, which was done, as it allows a nearly perfect match of the observed c_1 and Geneva $(U - B)$ indices, which are off by 0.06 and 0.04 mag, if a $\log g$ of 3.3 dex is assumed instead. However, there is no further overall improvement to spectrophotometry from using a model with individual chemical composition for this star.

In conclusion, the new model atmospheres with individual chemical composition are capable to reproduce colours and spectrophotometry of each of these four very cool CP2 stars rather well. Remaining discrepancies are reduced to either the IR and/or to a few parts of some of the flux features in the visual. But more importantly, the flux depression at 5200Å is represented much more accurately by these models than by models with

Table 20: Element abundances (in dex) from Lyubimkov (1986) for HD 196502 relative to solar values as in Table 17.

Element		Element		Element	
C	−0.50	Cr	+1.52	La	+0.67
N	−0.50	Mn	+0.37	Ce	+0.77
O	−0.50	Fe	+0.16	Pr	+1.48
Si	+0.01	Co	+0.44	Eu	+2.65
Sc	−1.25	Ni	+0.58	Gd	+1.44
Ti	+0.33	Y	+0.04		
V	−0.04	Zr	+0.75		

a scaled solar composition, as can easily be seen from Figure 41.

14.4 CP2 stars with $8500 \text{ K} \leqslant T_{\text{eff}} \leqslant 10000 \text{ K}$

To complete the sample of cool CP2 stars two objects were chosen which have been assigned the SrCr subtype, though they are known to have strong Eu features as well: HR 7575 (HD 188041) and 73 Dra (HD 196502). Both are slightly hotter than the four mCP stars already discussed and they show some of the largest known Δa values.

HR 7575 is particularly difficult to analyse with classical model atmospheres. Kato & Sadakane (1999) performed a thorough spectroscopic study of this star and found it impossible to achieve ionisation equilibrium for Cr and Fe with model atmospheres having the same T_{eff}, even when explicitly including Zeeman broadening in their radiative transfer computations. Remarkably, the same problem was also found for 73 Dra, while it was only of minor importance in β CrB (see their Table 5; the large magnetic field of β CrB requires to use a high pseudo–microturbulence, as it is done in the present analysis). For the initial model T_{eff}, pseudo-microturbulence, and element abundances was taken from the work of Kato & Sadakane (1999). CNO abundances were estimated by reducing the solar abundance value by −0.5 dex. But emerging spectrophotometric fluxes and model structure differ very little, if the averages from Roby & Lambert (1990) are used instead. This can be seen from a comparison between the change in fluxes for HD 204411 and HD 188041, as obtained from a scaled (+0.5 dex) solar CNO abundance and one based on the individual composition of the two stars (Table 17, Fig. 41 for HD 204411 and Fig. 42 for HD 188041) Note the carbon feature around 4750Å which was

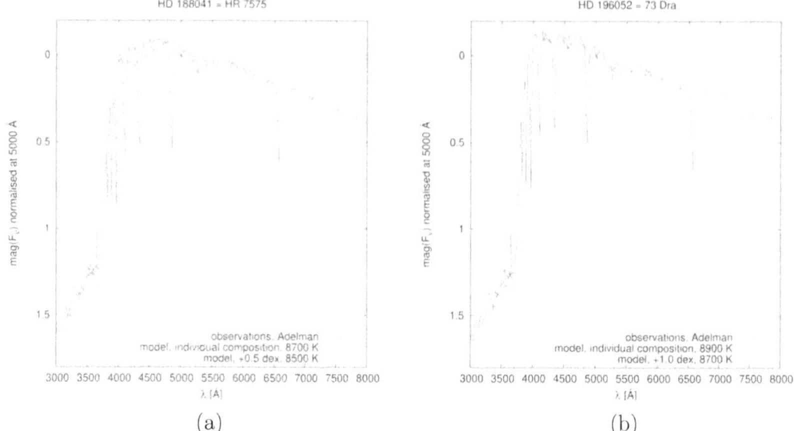

Figure 42: Left panel: spectrophotometric observations of HR 7575 are compared to fluxes from model atmospheres with an individual chemical composition and a model with enhanced metallicity (scaled solar abundances). Right panel: a similar comparison done for 73 Dra. Modelling and comparison details are discussed in the text.

also discussed in Smalley & Dworetsky (1993). One can conclude that the detailed choice of the CNO abundances does not significantly affect the overall flux distribution. For all other elements a solar abundance scaled by +0.5 dex was assumed and thus following Ryabchikova et al. (1999a) rather than Kato & Sadakane (1999), who had used +1.0 dex. The reason for this choice is that except for Cr, Mn, and the analysed lanthanides, which had been individually adjusted, the bulk opacity contributors, including Fe, were found overabundant by less than +1.0 dex in their own work. The suggested $T_{\rm eff}$ puts the star in the temperature range around the Balmer maximum, where a self-consistent derivation of both $T_{\rm eff}$ and $\log g$ from Balmer lines without knowing one of them accurately from an independent method is unfortunately unreliable even for "chemically normal" stars (Smalley et al., 2002). The discrepant ionisation temperatures, on the other hand, indicate basic shortcomings of chemically homogeneous atmospheres with scaled solar abundances, when they are used for the analysis of extreme CP stars. It was thus chosen $\log g$ so as to provide spectrophotometric fluxes in best overall agreement with observations.

For the same reason, the model atmosphere with scaled solar abundances used by Ryabchikova et al. (1999a) was selected as reference. Its basic parameters are given in Table 19. Fluxes from this model are compared to the data in Figure 42. Even with individual chemical composition a very high $\log g$ is required to match the Balmer jump and 4.5 dex appears to be a lower limit from this comparison. A T_{eff} of about 8700 K was preferable rather than 8500 K as used for the scaled solar abundant model atmosphere. Assuming either lower or higher temperatures makes the fit poorer in various wavelength regions, nor does it resolve the remaining discrepancies around 4200Å and 5200Å. With a T_{eff} of 8700 K the model based on an individual chemical composition also matches the observed fluxes in the red and IR region very well. Hence, the visual region fluxes of HR 7575 cannot be modelled as satisfactorily as those of the four cooler CP2 stars discussed above, although the predictions for the flux depressions around 4200Å and around 5200Å are a considerable improvement compared to fluxes from models with a scaled solar composition.

For 73 Dra, the initial parameters for the model atmospheres for this star were taken from the work of Ryabchikova et al. (1999a) whenever possible, as the spectroscopic analysis by Lyubimkov (1986) was based on outdated atomic data and model atmospheres. In particular, we followed their choice of T_{eff}, $\log g$, pseudo-mircoturbulence, and also assumed a solar element abundance scaled by +1.0 dex for all elements, for which no other determinations were available (Tables 16 and 19). Abundances for Cr, Fe, and Eu have been taken from Ryabchikova et al. (1999a). Otherwise, the abundances published by Lyubimkov (1986, and Table 20) were used, except for Ca. The very large underabundance he found is based on a single resonance line and appears unrealistic. Hence this result was substituted with a value of −0.5 dex below the solar Ca abundance as an indication of a suspected underabundance of Ca (solar abundance or even overabundance is the common case found for CP2 stars. CNO abundances were chosen as in the case of HR 7575.

As can be seen from Table 19, it was decided to use a slightly higher $\log g$ as well as a T_{eff} of 8900 K instead of 8700 K, which had been chosen for the model with scaled solar abundance. The reasons for this new choice are the following: it is possible to match the near IR fluxes with a model based on individual element abundances and a T_{eff} of 8500 K. However, the match in the blue and the UV achieved by the 8900 K model is lost in this case, and the remaining mismatch around the 5200Å feature becomes worse. Similar conclusions hold for the intermediate T_{eff} of 8700 K. The fundamental parameters thus cannot be adjusted to give a better match than that one shown in Fig. 42, even if a change in surface gravity is used to optimise the fit, as is indicated by the slightly higher value chosen for the final model. The mere scaling of a solar chemical composition does not reproduce the flux depressions in the 4200Å and 4600Å regions as well as the new model does. The flux predictions are also considerably improved for the 5200Å feature. This becomes evident when comparing the rather mediocre Δa value of the scaled solar abundance model with observations (Sect. 14.5). Thus, the new model atmosphere for 73 Dra is an improvement over its precursor, despite it cannot match the observations within the uncertainties of the data.

14.5 Comparison of the synthetic and observed Δa photometry for ten CP stars

In this Section, the synthetic Δa values with observations for the selected members of the λ Bootis, the CP1 and the CP2 group are compared. Table 21 lists all observed Δa indices along with those calculated from the new model atmospheres. One can conclude that the tendencies of the observed Δa values are well reproduced for all programme stars, if the limitations discussed in Sects 14.5.3 and 14.8 are taken into account. It also has to be emphasised that almost all magnetic CP stars exhibit photometric variability on the time basis of their rotational period (Sect. 5). As studies of the variation of element abundances of the target stars over their rotational period have not been available, it is not possible to give here a quantitative estimate of this effect. To provide a hint of its magnitude the range of observed Δa values in Table 21 published for two target stars are included. The same effect is even more prominent among hot CP2 stars.

14.5.1 λ Bootis stars

Members of the λ Bootis group exhibit significant negative Δa values (Sect. 11.7). Both programme stars show extreme underabundances (up to -2 dex) which results in Δa values of -13 and -20 mmag for 35 Aql and 29 Cyg, respectively. The synthetic values are about 5 mmag higher than the observed ones which is a satisfactory agreement. Notice that both objects are δ Scuti type pulsators (Paunzen et al., 2002a). The Δa values predicted by models with individual abundances and by those with an appropriately scaled solar element abundance are essentially the same for both stars. Heiter et al. (1998, Figure 4) presented a comparison of synthetic fluxes which includes the wavelength region covered by the Δa system.

14.5.2 CP1 stars

Only the most extreme CP1 stars have Δa values up to +14 mmag. In general, members of this group show only slightly enhanced values (Sect. 11.3). Photometrically, measurements always suffer from the fact that almost all CP1 stars are within spectroscopic binary systems. The two programme stars both have observed Δa values of +3 mmag whereas the synthetic ones are +8 mmag. Again this is a remarkably good agreement showing that the 5200Å flux depression is only marginally pronounced in CP1 stars. Model atmospheres with a scaled solar element abundance increase the synthetic Δa values to about $+10 \cdots + 12$ mmag. This is essentially due to the increased microturbulence of 4 km s^{-1} used in all the model calculations for these two stars, despite the fact that such a value is even below actual microturbulence velocities determined for either of them (Table 18 and compare with Table 16). The CP1 stars hence represent quite the opposite to the case of magnetic CP stars discussed below, because large values of ξ_t which are based on abundance analyses lead to an overestimate of Δa when combined with scaled solar element abundances. The discrepancy is reduced to an acceptable amount, if the model atmospheres are based on individual element abundances. For the CP2 stars exactly the contrary was found, as shown in the following.

14.5.3 CP2 stars

The magnetic CP2 stars are the most investigated within the Δa photometric systems. The measured values can go up to +75 mmag and vary over the rotational period. For the four cooler programme stars (HD 137909, HD 137949, HD 201601, and HD 204411) a good agreement of the measured (+11 to +26 mmag) and calculated values was found. This is even more remarkable since these objects have strong surface magnetic fields from 0.5 to 5.7 kG (Table 16). In contrast, the Δa values predicted by the best model atmospheres that can be constructed with scaled solar element abundances, as discussed in Sect. 14.3.2 and summarised in Table 19, are only +12 mmag for HD 137909 and HD 137949 as well as +2 and +3 mmag for HD 201601 and HD 204411. The flux depression at 5200 Å is hence severely underestimated by these models, in accordance with the comparison to spectrophotometry in Section 14.3.2. The slightly higher Δa values of +12 mmag obtained for model atmospheres for β CrB and 33 Lib with a scaled solar metallicity are entirely due to an enhanced (pseudo-) microturbulence used in their computation (Table 16). The beneficial effect of increasing this parameter has already been pointed out by Adelman &

Table 21: Δa indices in mmag for the target stars. The values of $T_{\rm eff}$ and $(b-y)_0$ given are those calculated from the best fitting models from Sect. 14.3 which were also used to determine the calculated Δa given in the fifth column. They are compared with observations of Δa listed in the fourth column which is based on data taken Section 10. Ranges in Δa obs. are examples for variations over the rotational period, ranges in Δa calc. indicate typical model atmosphere parameter uncertainties.

HD	$T_{\rm eff}$ [K]	$(b-y)_0$ [mag]	Δa obs. [mmag]	Δa calc. [mmag]
108642	8200	0.068	+3	$+7\cdots+8$
108651	7900	0.106	+3	+8
137909	7750	0.142	+26	$+29\cdots+33$
137949	7550	0.205	$+20\cdots+25$	$+34\cdots+40$
183324	9300	0.017	−13	−8
188041	8700	0.057	$+59\cdots+90$	+22
192640	7800	0.139	−20	$-12\cdots-13$
196502	8900	0.010	+72	+32
201601	7700	0.130	+11	$+9\cdots+11$
204411	8500	0.036	+19	+10

Rayle (2000). However, this increase generally is not sufficient to reproduce the observed Δa values and a larger variation would have to be justified by extraordinary magnetic field strengths. Moreover, in CP1 stars the same values of enhanced microturbulence velocities are used for a completely different physical reason, yet leading ultimately to the same Δa value when using a model atmosphere with scaled solar metallicity. The observed Δa values of CP1 stars, however, are much lower than those of CP2 stars. On the other hand, combining an enhanced microturbulence with individual abundances for cool CP2 stars brings the observed Δa values easily within the range of synthetic calculations through enhancing them even further, while, as discussed before, those of CP1 stars are reduced in comparison with the case of a scaled solar abundance. This is confirmed by the results shown in Table 21.

For the two hotter objects, HD 188041 and HD 196502, the synthetic Δa values are about +40 mmag too low. Both stars are photometrically variable with a very high amplitude (Catalano et al., 1998a) which could account for some part of the underestimation of the 5200Å flux depression when using these models. The models with scaled solar abundance again show a much poorer performance. A value of +10 mmag is found for HD 188041 only if an enhanced (pseudo-) microturbulence of 4 km s^{-1} is used. The model with individual element abundance assumes a value of 1.75 km s^{-1} (Table 16) as obtained by the analysis of Kato & Sadakane (1999). For 73 Dra the comparison is more straightforward, because all model calculations for this star have been performed with 2 km s^{-1} (Table 16). In that case, despite an overabundance of +1.0 dex, the model with scaled solar element abundance only yields a pathetic value of +6 mmag compared to +32 mmag obtained from the model with individual element abundances, which has to be compared to the +72 mmag found in observations.

Therefore, the two hotter CP2 stars in the present sample demonstrate even more clearly the improvement obtained from using individual element abundances for modelling the 5200Å flux depression. Usage of an enhanced (pseudo-) microturbulence could yield an even closer agreement, but it is also a reminder that a realistic modelling of the magnetic field of these stars is yet a missing ingredient of even the most recent model atmospheres. To the role of stratification for these matters is discussed in Section 14.8.

14.5.4 Spectroscopic observations of the 5200Å depression

In this Section, a comparison of observed high resolution spectra with synthetic ones is presented. It is not intended to check the validity of the abundances used in all detail but show that it is possible to fit the average line blanketing in the 5200Å depression within acceptable limits using the described models. This corroborates the direct comparison of synthetic and observed Δa values as well as synthetic and observed spectrophotometry given in the previous sections.

Observations of all programme stars were made with the coudé spectrograph at the 2-m RCC telescope of the Bulgarian National Astronomical Observatory Rozhen during 2000 and 2001 by Ilian Kh. Iliev. The Photometrics AT200 camera with back-illuminated SITe SI003AB (1024 × 1024, 24 μ pixel) CCD chip attached to the Third camera was used. The spectra were taken with a 632 mm^{-1} Bausch & Lomb grating resulting in a resolution of about 20000. Due to the "classical" coudé scheme the entire spectral range between 5000Å and 5400Å was covered by eight consecutive and fairly overlapping sub-regions. Each of these sub-regions is 100Å large.

A hollow-cathode ThAr lamp was used to produce a reference spectrum with a FWHM of about 2 pixels for the comparison lines. Flat fields were made by using a tungsten projection lamp mounted in front of the entrance slit. The spectra were bias-subtracted and flat-fielded within standard IRAF procedures. The typical signal-to-noise ratio between 200 and 300 in the continuum was achieved for most of the spectra. Wavelength calibration resulted in an r.m.s. wavelength error of typically 0.005Å.

Figure 43 shows the difference of the observed and synthetic spectrum for two CP2 stars (HD 137909 and HD 204411) which have the largest and smallest scatter, respectively. The other "differential" spectra are very similar and are "in between" the cases shown here. From this figure it can be concluded that most of the strong lines in the region have been included in the computations. In many cases the same lines (e.g. around 5170Å) deviate which is quite likely caused by wrong or incomplete atomic line data. Otherwise, the line blanketing at the 5200Å depression is well reproduced with a scatter of 5 to 10% of the differential spectra around a (nearly) zero mean.

14.6 Observational constraints

If a model atmosphere would be at hand for which the overall emergent surface flux distribution matches the observed one at the few percent level, as it is possible for example for Vega (Castelli & Kurucz, 1994), the Δa indices derived within both observed and synthetic systems should agree to within a few mmag (Sect. 14.1.2). This should hold unless flux deviations in the wavelength range of the Δa filters were larger than average, for instance due to a peculiarity in the radiative transfer of the stellar atmosphere which has not been accounted for by the model. The results of Sections 14.3 and 14.5 show that this is indeed what happens for models of CP1 and CP2 stars, if a scaled solar element abundance is assumed. The comparison of observed spectrophotometry with calculations from the models with individual abundances, however, reveals a better agreement not only over most of the visual region, but in particular for the wavelength range around 5200Å (Figs. 39 to 41). Table 21 shows that observed and synthetic Δa values obtained from the new models frequently agree to within a few mmag. Even the two hotter CP2 target stars, HR 7575 and 73 Dra, fit within this framework. However, it is already obvious from Fig. 42 that despite an overall agreement of observed and synthetic fluxes in the visual region, the flux depression at 5200Å is underestimated by all models available for these two stars, and indeed the Δa values listed in Table 21 agree with this expectation. Hence, comparing observational and synthetic Δa values is a reliable method to judge whether a model atmosphere for an individual CP star is capable to predict the extent of a flux depression at 5200Å or not.

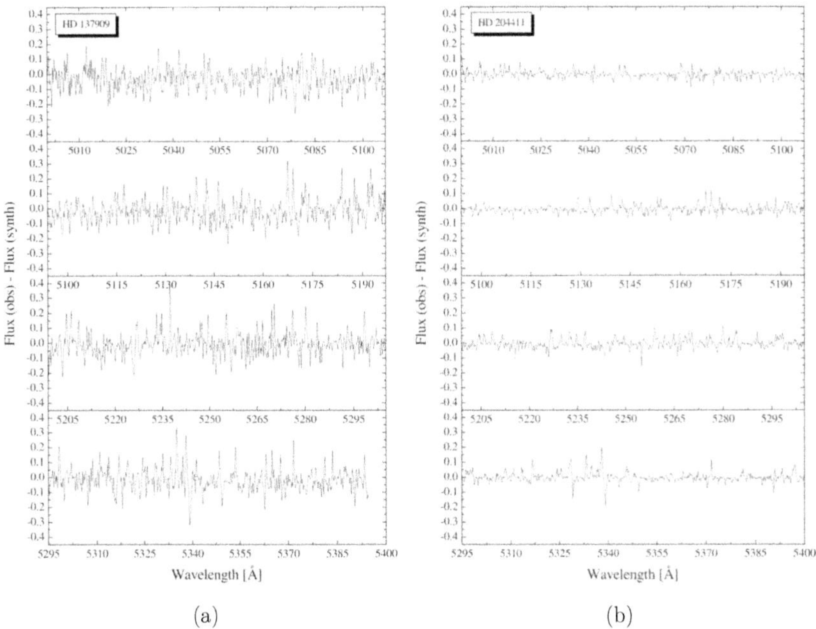

Figure 43: A comparison of high resolution observed and synthetic spectra for HD 137909 and HD 204411. Note that the noise level of both sets of observations is roughly the same, hence the larger deviations for HD 137909 are significant. They may indicate missing weaker lines in the data base, but HD 137909 also has a much stronger magnetic field and is known to be vertically stratified (see text in Sects. 14.5.4 and 14.8 for further discussion). Fluxes as used here are relative to local continuum contrary to the much lower resolution spectrophotometry shown in Figure 41.

The accuracy of this assertion is based on how well one knows the spectrophotometric flux distribution of a target star, as this method was chosen to determine the stellar fundamental parameters. A variation of $T_{\rm eff}$ by ± 200 K for 73 Dra changes Δa by about 2 to 3 mmag, while a variation of $\log g$ by ± 0.2 dex changes it by just 1 mmag. To distinguish between these internal errors of parameter determination and those which can be achieved on an absolute scale would require a better knowledge of the effects of atmospheric extinction and instrumental calibration on the spectrophotometric data. As far as atmospheric extinction and instrumental effects are concerned, no alternative but to accept the limitations of the current data is available. Problems with data normalisation perhaps account for some of the deviations noted in the far red and IR for most of the target stars in Section 14.3. For that reason less weight is given to those differences. Notably, the same phenomenon was found for the standard model atmosphere of Vega published by Castelli & Kurucz (1994). From their Figures 1 and 5 it is clear that fluxes of ATLAS9 and ATLAS12 models are too low redward of H_α. This problem could also be caused by the flux calibration of Vega rather than by an intrinsic shortcoming of the model atmospheres. Hence, a recalibration of our standard sources as well as a new generation of spectrophotometric flux measurements would be a major step towards tighter error margins in the determination of fundamental parameters through spectrophotometry.

14.7 Microturbulence, abundances, and atomic data

Although the microturbulent velocity ξ_t is usually associated with a photospheric velocity field, it is also used to mimic enhanced line blanketing due to Zeeman splitting in a magnetic field (Sect. 14.2.1). One of the results of Sect. 14.5 is that an increase of ξ_t from 2 to 4 km s^{-1} can raise synthetic Δa values by up to +10 mmag. From an enhanced ξ_t alone it is not possible to distinguish between CP1 and CP2 stars, because they may be modelled with similar values of this parameter, but for different physical reasons. Model atmospheres with scaled solar element abundance leave this distinction between CP1 and CP2 stars unexplained. As shown in Sect. 14.5, they yield values for Δa which are too large for CP1 stars, while for many of the CP2 stars which have a Δa of +20 mmag or more the increase provided by an already large ξ_t of 4 km s^{-1} is not sufficient. This explains why the previous model atmospheres based on grid calculations with pretabulated line opacities and fixed, scaled solar element abundances cannot be used to analyse the contributors to the flux depression in the 5200Å region, although the latter can sometimes be matched by such models, as noted in Adelman & Rayle (2000). Only a combination of both individual element abundances and ξ_t, whether the latter be attributed to a velocity field as in CP1 stars or to a magnetic one as in CP2 stars, recovers the correct distinction between the two: mostly marginal Δa values for CP1 stars and moderate to large ones for CP2 stars. Consequently, both the effect of magnetic fields (Zeeman splitting) and the particular abundance pattern are required to understand the large flux depressions around 5200Å found for many CP2 stars. On the same grounds it can be also understood why some extreme CP1 stars have moderate Δa values of up to +14 mmag (which are also found in some CP2 stars), despite the fact that they do not show detectable magnetic fields.

This leaves one with uncertainties in element abundances and thus in the amount of line opacity contributing to the 5200Å flux depression as a problem to look into for judging the new models. Derived element abundances depend on the model atmosphere used, and hence on fundamental parameters assumed, on the atomic data taken as input for model construction and radiative transfer calculations, and, particularly for the CP2 stars, also on magnetic field and stratification effects not accounted for in the present models. The accuracy and completeness of atomic line data affects any approach to construct model atmospheres for CP stars. Examples relevant for CP2 stars were given by Cowley et al. (2000). They had to artificially increase iron peak element abundances to compensate for missing line opacities due to lanthanides in their model atmosphere for the extreme, very cool CP2 star HD 101065, and it was argued (Wahlgren et al., 2001) that the third spectra of lanthanides may contribute to the flux depression at 5200Å. However, no major improvement over the present results from just more complete atomic data alone is expected. The figures in Sect. 14.3 and Table 21 underline the success of the new models to predict emergent fluxes from the UV to the red and particularly in the 5200Å region. The lines of Nd III discussed by Wahlgren et al. (2001) have already been included in the calculations as they are part of VALD. Note that while it is not possible to exclude the presence of autoionisation lines, the results from Sects. 14.3 and 14.5 indicate that they do not seem likely to play any role for the 5200Å flux depression for the CP stars with a $T_{\rm eff}$ of less than 10 000 K.

14.8 Stratification and magnetic fields

For the model atmospheres a homogeneous composition was assumed and also a solar abundance of He. The present studies of stratification effects are based on phenomenological multi-layer models (Cowley et al., 2009) which differ from the approach used here only by assuming a different average composition in two or three vertical layers (sometimes with smooth transitions zones in between) which build up the photosphere. This setup allows a much closer match of the observed spectral line profiles of cool Ap stars and is supported by their gross agreement with diffusion calculations. But the latter in turn are not based on a model atmosphere with a self-consistent treatment of magneto–convection (see below). Further research is still needed to corroborate the conclusions from these more refined models of atmospheres of magnetic Ap stars.

Here, an enhanced microturbulence was used to model the effect of line blanketing in strong magnetic fields on average surface fluxes, because a direct modelling of a global stellar magnetic field (and ultimately a global, three dimensional stellar model atmosphere) in this context is so far not possible. The analytical and numerical complexity of a model stellar atmosphere are considerably increased and suitable (observational) data are required to constrain the topology and strength of the field. Even with these data at hands, it is not expect that they be sufficient on their own, as vertical stratification of element abundances may easily influence observed fluxes by an even larger amount than a strong magnetic field (Alecian & Stift, 2010).

For the cool CP2 stars an additional physical problem is introduced by the convective instability of the H ionisation zone situated in their lower photosphere. Vertical stratification models (and diffusion calculations which provide their theoretical justification) have not yet been coupled to magneto-convection in a self-consistent manner, owing to the complexity of the problem, despite the unforseeable changes of the results one might expect for that case. For instance, contrary to common belief, a magnetic field does not always have to suppress convection. One example for such a scenario is normal solar granulation (as in magnetically quiet regions – not to be confused with what happens in sunspots) for which, on the basis of more refined measurement techniques available, quite considerable field strengths are thought to be present (López Ariste & Sainz Dalda, 2012). These obviously cannot suppress solar convection. For a more simplified physical problem, Canuto & Hartke (1986) have shown that depending on the field strength and efficiency of radiative losses even the opposite may occur: at weak to moderate field strengths the convective flux can be enhanced for a certain range of Rayleigh numbers (and hence radiative loss rates) because of an extra contribution originating from the transport of magnetic energy. At larger field strengths the flux may indeed be suppressed for the case of inefficient convection (with small to moderate Rayleigh numbers), as has already been shown much earlier for several cases by Gough & Tayler (1966). However, for very efficient convection, even a strong field rather results in a change of shape and orientation of convective cells. More accurate predictions could be expected from magneto–hydrodynamical simulations. However, to obtain quantitative results for individual cool CP2 stars from this technique requires to account for a magnetic field which changes on the scale of the entire stellar surface and which is coupled to the local velocity fields that have to be resolved at least on the scale of up- and downflows by a hydrodynamical simulation. Evidently, such simulations pose a serious, though interesting computational challenge.

14.9 Stellar model atmospheres with magnetic line blanketing

In a series of three papers, Kochukhov et al. (2005); Khan & Shulyak (2006a,b), developed new model atmospheres of A- and B-type taking into account magnetic line blanketing and polarized radiative transfer. These calculations are based on the new stellar model atmosphere code LLMODELS (Sect. 14.9.1) which implements direct treatment of the opacities due to the bound-bound transitions. The anomalous Zeeman effect is calculated for the field strengths between 1 and 40 kG and a field vector perpendicular to the line of sight. The model structure, high-resolution energy distribution, photometric colours, metallic line spectra and the hydrogen Balmer line profiles were computed for magnetic stars with different metallicities and were discussed with respect to those of non-magnetic reference models. The magnetically enhanced line blanketing changes the atmospheric structure and leads to a redistribution of energy in the stellar spectrum.

They also studied the effects of the magnetic field, varying its strength and orientation, on the model atmosphere structure, the energy distribution, photometric colours and the hydrogen Balmer line profiles. A probable explanation of the visual flux depressions (especially at 5200Å) of the magnetic CP stars in the context of these models was presented. They calculated a grid of model atmospheres of magnetic A- and B-type stars for different effective temperatures ($T_{\rm eff}$ = 8000, 11000, and 15000 K), magnetic field strengths (B = 0, 5, 10, 40 kG) and various angles of the magnetic field (Ω = 0 - 90°).

The models, techniques, and results are presented here because they investigated the behaviour of the Δa, $(\Delta V1 - G)$, and Z (Sect. 4) depending on different metallicities and magnetic field configurations. This can be seen as supplement and update to the models as well as results presented in Sections 14.1.2 and 14.2.

14.9.1 The stellar model atmosphere code LLModels

The stellar model atmosphere code LLMODELS developed by Shulyak et al. (2004) uses a direct method, the so-called line-by-line or LL technique, for the line opacity calculation. Such an approach allows to account for the anomalous Zeeman splitting of spectral lines in the line blanketing calculation and, hence, to achieve a qualitatively better level of accuracy in modelling the influence of the magnetic field on the stellar atmospheric structure. LLMODELS aims to model early and intermediate type stars taking into account their individual chemical composition and an inhomogeneous vertical distribution of elemental abundances. The code is based on the modified ATLAS9 subroutines (Kurucz, 1993) and on the spectrum synthesis code described by Tsymbal (1996). It uses the following general approximations:

- a plane-parallel geometry is assumed;
- Local Thermodynamic Equilibrium (LTE) is used to calculate the atomic level populations for all chemical species;
- the stellar atmosphere is assumed to be in a steady state;
- the radiative equilibrium condition is fulfilled.

The main goal of the LL modelling technique is to avoid a simplified or statistical description of the bound-bound opacity in stellar atmospheres. The temperature distribution for a given model strongly depends upon the accuracy of the flux integration over the whole spectral range where the star radiates significantly. Consequently, to ensure an accurate flux integration, the number of frequency points has to be sufficient to provide a realistic description of individual spectral lines. In the past, limited computer resources and imperfect numerical techniques were applied to solve this problem. The Opacity Distribution Function method (ODF, Kurucz, 1979) and the Opacity Sampling technique (OS, Gustafsson et al., 1976) are employed. Within the LL method, no precalculated opacity tables are needed. Furthermore, no approximations about the line opacity coefficient are made during the model calculation. By using a large number of wavelength points (up to 500 000), a detailed description of the line absorption for all atmospheric depths and achieve a higher dynamical range in opacity is provided. This allows to reach the required accuracy of the resulting model atmosphere structure, especially in the upper atmospheric layers. The atomic line data provided by Kurucz (1992) and the VALD project (Kupka et al., 1999) are used.

For the detailed implementation of the Zeeman effect and the magnetic field configuration, the reader is referred to Kochukhov et al. (2005); Khan & Shulyak (2006a,b).

14.9.2 The behaviour of Δa, $(\Delta V1 - G)$, and Z in correlation with T_{eff} and B

Here, a summary of the results for Δa, $(\Delta V1 - G)$, and Z depending on different metallicities and magnetic field configurations is given.

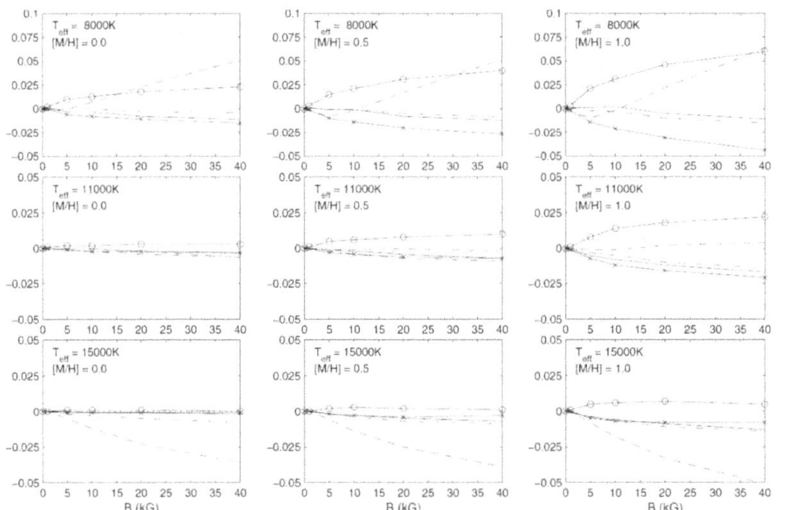

Figure 44: Difference between photometric colour indices of magnetic stars in comparison with the predictions for the non-magnetic reference models taken from Kochukhov et al. (2005). The effect of magnetic field on the photometric parameters of the Strömgren system is shown for $(b-y)$ (solid line), m_1 (dotted line), c_1 (dashed line) and β (dash-dotted line). The peculiarity indicators in the Geneva 7-colour system, $(\Delta V1 - G)$ and Z, are shown with open circles and crosses, respectively.

Kochukhov et al. (2005); Khan & Shulyak (2006a) studied the influence of magnetic line blanketing on the different photometric systems, besides Δa, also Strömgren $uvby\beta$ and the Geneva 7-colour system. All colours were calculated using modified computer codes by Kurucz (1993), which take into account transmission curves of individual photometric filters, mirror reflectivity and a photomultiplier response function. The synthetic colours are computed from the energy distributions sampled every 0.1Å, so integration errors are expected to be negligible.

Having introduced a more accurate treatment of Zeeman splitting on line blanketing, it is particular interesting to study specific photometric indices (Δa, $(\Delta V1 - G)$, and Z) whose aim is to identify CP stars. The behaviour of the photometric indices is closely related to the flux distribution between the visual and UV regions (Sect. 14.3) and the presence of the flux depression at 5200Å.

A relation between changes of all photometric indices and the intensity of magnetic field depends strongly on the stellar effective temperature.

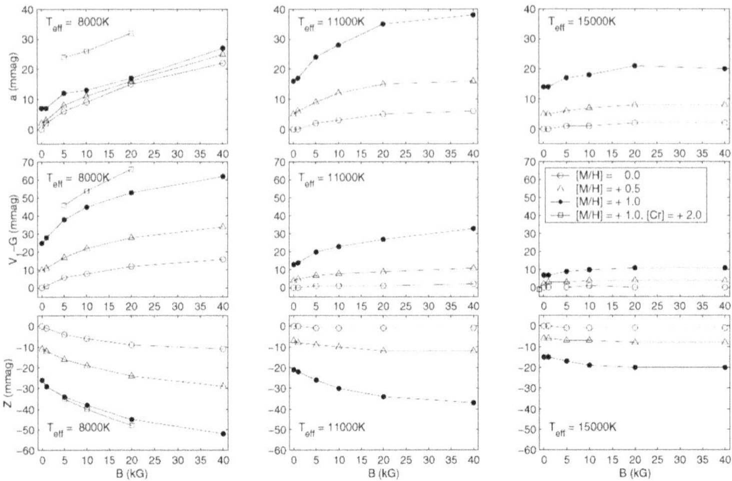

Figure 45: The magnitude of the a, $V1-G$ and Z photometric indices as a function of the magnetic field strength for $T_{\text{eff}} = 8000$, 11000 and 15000 K taken from Khan & Shulyak (2006a). Different curves show calculations with metal abundances [M/H] = 0.0 (open circles), [M/H] = +0.5 (triangles) and [M/H] = +1.0 (filled circles). Open squares illustrate the effect of increasing Cr overabundance to [Cr] = +2.0 in the $T_{\text{eff}} = 8000$ K, [M/H] = +1.0 models. All curves are shifted so that for [M/H] = 0.0 and zero magnetic field in each frame the a, $V1 - G$ and Z indices equal zero.

For low T_{eff}, photometric changes are very pronounced, whereas for hotter magnetic stars modification of the $uvby\beta$ photometric observables is fairly small (except c_1). The Geneva photometric colours show a similar decrease of the sensitivity to the magnetic effects with increasing T_{eff}. These effects are shown in Figure 44.

The results of the calculations of the synthetic a, $(V1 - G)$, and Z parameters by Khan & Shulyak (2006a) are shown in Figure 45.

The Δa system is most sensitive to the magnetic field strength, except for model atmospheres with $T_{\text{eff}} = 8000$ K where the $(\Delta V1 - G)$ demonstrates the best sensitivity. As already shown in Sect. 14.2, it is not very sensitive to metallicity (although it does increase with increasing B) for this temperature range. Khan & Shulyak (2006a) supposes that this phenomena appears due to some saturation effect of enhanced line blanketing in the narrow region around 5200Å, which is mainly dominated by the Fe I and low excitation Fe II lines for low effective temperature. In contrast to the narrow-band Δa-system, the sensitivity of medium-band parameters $(V1 - G)$ and Z does not suffer much from the specific properties of spectral feature in the narrow interval around 5200Å. This state is well illustrated in flux units (Fig. 46). One can see that the flux depression in the region longward of 5200Å is clearly affected by additional abundance and magnetic intensification, while the narrow region centred at 5200Å is not altered much.

To test the influence of an anomalous concentration of an individual species on the photometric peculiarity parameters, and especially the Δa value for low effective temperature, Khan & Shulyak (2006a) increased the Cr overabundance by a factor 100 relative to the Sun and calculated additional model atmospheres. As illustrated in Figs. 45 and 46, this results in 12 – 15 mmag growth of the Δa index for $T_{\text{eff}} = 8000$ K, [M/H] = +1.0 model and field strength 5 – 20 kG. For $(\Delta V1 - G)$ the increase is smaller and amounts to 8 – 13 mmag while, the Z-index is almost unaffected (1 – 3 mmag). This behaviour indicates that observable anomalous Δa values for low effective temperature may arise not only from the influence of the magnetic field but also from specific anomalous abundances as well. Nevertheless, even with overabundant Cr, the Δa system demonstrates the least sensitivity to the magnetic field strength, with only 8 mmag change from 5 – 20 kG, while the value for Z is 13 mmag and 20 mmag for $(\Delta V1 - G)$.

Finally, none of the photometric peculiarity indicators (except a for $T_{\text{eff}} = 8000$ K) show a linear trend over the whole range of the considered magnetic field strength. Saturation effects for their changes with the magnetic field intensity appear for $B \gtrsim 10$ kG. The Δa index for low effective temperature ($T_{\text{eff}} = 8000$ K) demonstrates almost linear dependency on B from 5 to 40 kG.

As a conclusion it can be said, that the Δa photometric system is best suitable to detect not only magnetic fields but also CP stars of the upper MS.

14.9.3 Effects of the individual abundance patterns

As last part of their detailed investigation of CP stars, Khan & Shulyak (2007) presented the effects of individual abundance patterns on the model atmospheres of CP stars which is summarized here. The main purpose was to conduct a systematic homogeneous study to explore the abundance parameter space occupied by these stars. They calculated a grid of the model atmospheres for B- and A-type MS stars for T_{eff} values between

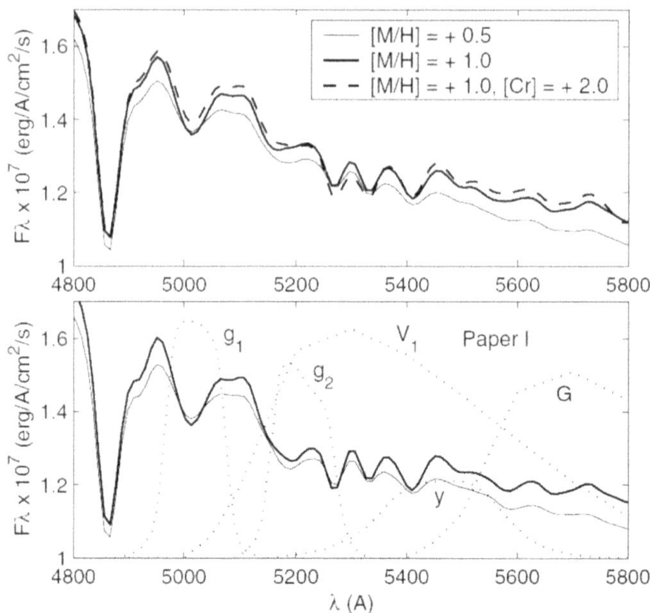

Figure 46: Energy distribution around 5200Å for $T_{\rm eff} = 8000\,{\rm K}$, $B = 20\,{\rm kG}$ and two values of metallicity $[{\rm M/H}] = +0.5$ and $[{\rm M/H}] = +1.0$; for the second metallicity value, the behaviour for overabundant $[{\rm Cr}] = +2.0$ is also shown. The top frame represents the comparison of fluxes from Kochukhov et al. (2005), the bottom frame shows the respective data from Khan & Shulyak (2006a), including polarized radiative transfer. The fluxes are convolved with a Gaussian profile assuming a resolution $R = 150$. The curves of photometric filters (dash-dotted lines) are presented for comparison. The g_1, g_2, and y are profiles of filters of the Δa system, and $V1$ and G of the Geneva 7-colour system.

20 000 and 8000 K, and chemical compositions. Using the LLmodels codes (Sect. 14.9.1) to compute model atmospheres with individual abundance patterns, the following elements were varied: C, Mg, Si, Ca, Ti, Cr, Mn, Fe, Ni, Sr, Eu, and He. They presented a homogeneous study of model atmosphere temperature structure, energy distribution, and photometric indices in the Strömgren $uvby\beta$ and Δa systems, hydrogen line profiles, and the abundance determination procedure as it applies to CP stars. The draw several conclusions (summarized in the following) which are important for the interpretation of the Δa index for several star groups (Sects. 6 and 7).

The majority of the tested chemical elements (within the limits of abundance values considered) produce less than 1% variations in the model atmosphere temperature profile and fall into the small changes group. These elements are He, C, N, O, Mg (deficient), Ca, Sr, Eu, and Hg. Several elements were assigned to the moderate changes group in accordance with the effects they produce on the model atmosphere structure (1 – 3%) and the energy distribution. These elements were Mn, Ni, and Mg (enhanced). The group of elements which produced large changes in the model atmosphere structure (more than 3%) and energy distribution consisted of Si, Cr, Fe and scaled abundance patterns.

If they considered changes in the temperature structure produced by the elements Si, Cr, and Fe in the main line forming region then they found that the sequence of Fe–Si–Cr represents these elements from the most to the least influential. Model atmospheres peculiar in Cr, Mn, and Fe demonstrate a very similar, highly organized, temperature behaviour. There were two distinct cooling and heating regions in the upper and the lower (i.e. the main line forming region) atmosphere, respectively. As the abundance value grew, the temperature dropped in the upper atmosphere and increased in the lower atmosphere. The inflection point which separates these two regions moves outwards with growing $T_{\rm eff}$. The magnitude of the heating region steadily grew with increasing $T_{\rm eff}$ excluding the lowest value of 8000 K. The same effects, except for the cooling region feature, applied to models with scaled abundances as well. The reason why Cr, Mn, and Fe demonstrated such a behaviour is likely the similarity of their energy level configurations (and the corresponding spectral line distribution patterns), and their relatively high content in the atmosphere.

They concluded that the elements Si and Fe are the main providers of two different types of the temperature changes to produce the same distinctive temperature structure that model atmospheres with scaled abundances do. In other words, the temperature structure of a model atmospheres peculiar only in Si and Fe may be close to the temperature profile of a models with scaled abundances. The agreement is quite good for low effective temperatures, while for $T_{\rm eff} > 11\,000$ K the cumulative effect of all other chemical elements was required. They suspected that overabundant C, Mg, and Ca are of the most significance.

They found that Fe is the principal contributor into the 5200Å depression for the whole range of $T_{\rm eff}$, while Cr and Si are important primarily for low $T_{\rm eff}$. The analysis of the diagram of the peculiarity index a versus $(b - y)$ for the model atmospheres peculiar in Si, Cr, and Fe revealed regular patterns in the locus of points representing those models. In fact, there were three separate directions (axes) associated with the growing content of each element on the diagram. The inclination of the Fe-axis to the normality line (Sect. 3) and the fact that it did not depend on $T_{\rm eff}$ clearly demonstrated that Fe is indeed the key element to the fundamental property of the Δa system.

15 Heuristic observational correlations for CP stars

The CP stars of the upper MS (Sect. 6) were the primarily targets of Δa observations over the last four decades in the Galactic field (Sect. 10), open clusters (Sect. 17), and the LMC (Sect. 19). Therefore, an overview of some heuristic observational correlations for CP stars are given. The theoretical models and synthetic photometry are described in Sects. 8.1, 9, and 14.

Abundances: Netopil (2013) investigated the correlation of Δa and $\Delta(V1-G)$ (Sect. 4) with the abundances in more detail. Here, these results are reviewed. Ryabchikova (2005) presented the observational evidence that the abundances of Fe and Cr are strongly depending on the $T_{\rm eff}$ which led to the conclusion that these are mainly responsible for the absolute value of Δa. Later, Khan & Shulyak (2007) confirmed this result on the basis of stellar atmospheres (Sect. 14.9.3). Netopil (2013) used the abundances listed in Ryabchikova (2005) as a starting point, and updated it with the recent literature. If more investigations were available for one object, an average of the abundances have been derived. In total, abundances of the major elements for 87 magnetic chemically peculiar objects were compiled. If averaged $T_{\rm eff}$ were available in the lists by Netopil et al. (2008), these were adopted, otherwise the values given in the respective references were used. From this sample, 50 stars have a measured Δa value (Sect. 10). The results are shown in Figure 47. Although due to the different sensitivity of these two peculiarity indices on specific spectral features (Sect. 14.9.2), a direct comparison is somehow problematic, an offset of +15 mmag was applied to $\Delta(V1-G)$ to obtain a comparable peculiar index for all objects. The upper panel shows the $T_{\rm eff}$ dependency of the photometric indices, whereas the others the dependency of selected elements. The correlation between the Fe abundance and the peculiar indices can be easily recognized also in Fig. 48, supporting the conclusion that Fe is the main contributor to the flux depression at 5200Å and hence for positive Δa values. Its "zero point" is located at about 7500 K, coinciding also with solar Fe abundances, whereas the respective solar reference values were taken from Asplund et al. (2009). The hotter end of the sample is

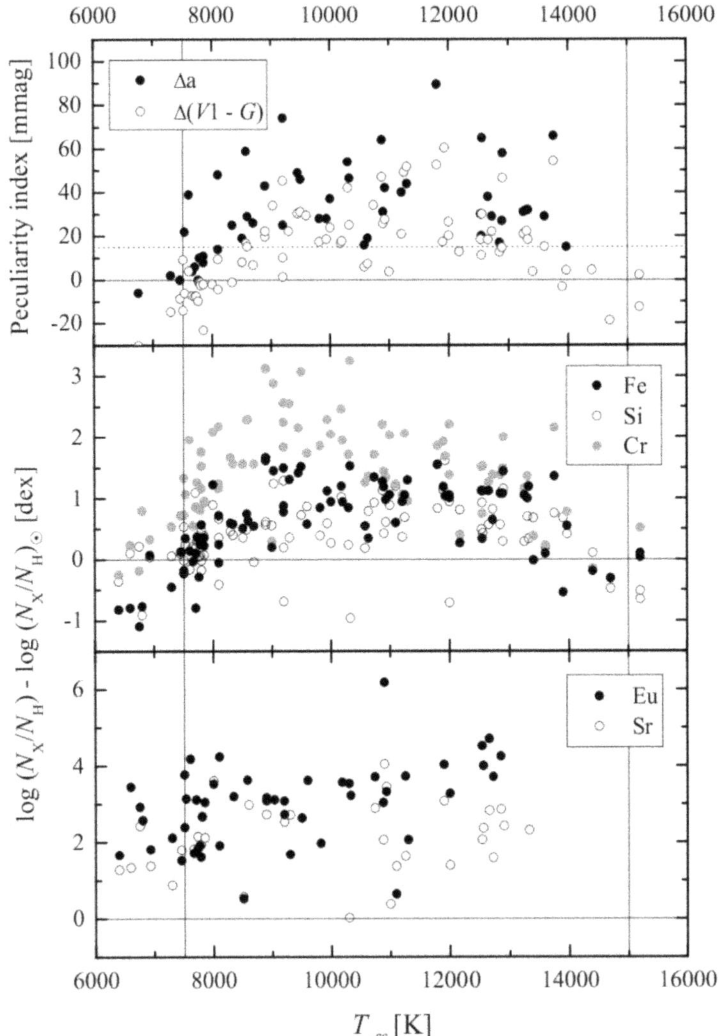

Figure 47: The T_{eff} dependency of Δa (filled circles) and $\Delta(V1-G)$ (open circles) is shown in the upper panel. The $\Delta(V1-G)$ values were shifted by +15 mmag. The solid lines indicate the zero point of Δa at 7400 K and the dotted line a Δa limit of +15 mmag. The lower two panels show the abundances of some selected elements versus T_{eff}. The data were taken from Netopil (2013).

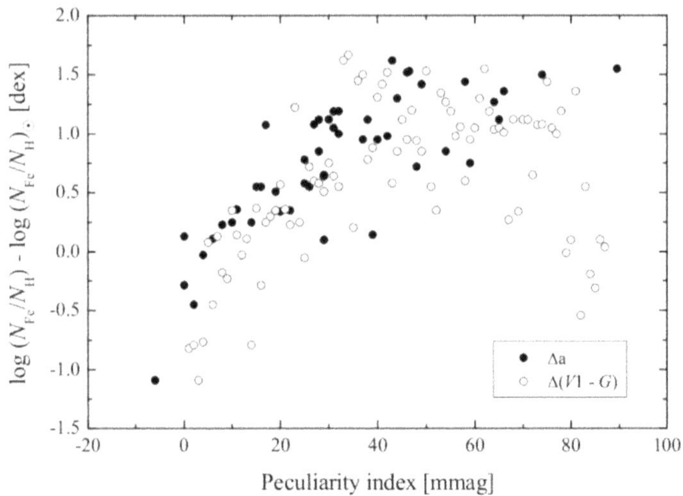

Figure 48: The [Fe/H] abundances of CP stars in correlation with Δa (filled circles) and $\Delta(V1 - G)$ (open circles).

not optimal covered, but is roughly estimated at about 15 000 K. However, Δa values can be designated as significant only at a level of +15 mmag, depending on the scatter of the normal type stars (Sect. 11.1). Hence, taking into account this significance level, the $T_{\rm eff}$ range decreases to about 8000 − 14 000 K. Cooler or hotter CP stars will be probably only detected if large overabundances of Si or Cr are present and/or a strong magnetic field contributes to enhance Δa. The overall lower Δa values at higher $T_{\rm eff}$ can be also noticed in the models by Khan & Shulyak (2006a). Nevertheless, in this $T_{\rm eff}$ range, all stars of the investigated sample (Fig. 47) would have been unambiguously detected with Δa photometry.

Colour: Figure 49 shows the Δa versus $(b - y)_0$ diagram for all stars (Sect. 10) included in Renson & Manfroid (2009, RM09) together with the detection limit of CP stars according to Section 11.1. The red circles mark the objects therein which are well-established CP stars of all types. These stars seem to be uniformly distributed over the complete colour range with an apparent gap at $0 < (b - y)_0 < +0.05$ mag. This is not only the region where the Balmer jump is at it maximum, but also where the estimation of the reddening is most difficult (Napiwotzki et al., 1993). However, although the estimation of the reddening for the sample has been cross checked with several methods (Sect. 10.1), there is small chance of

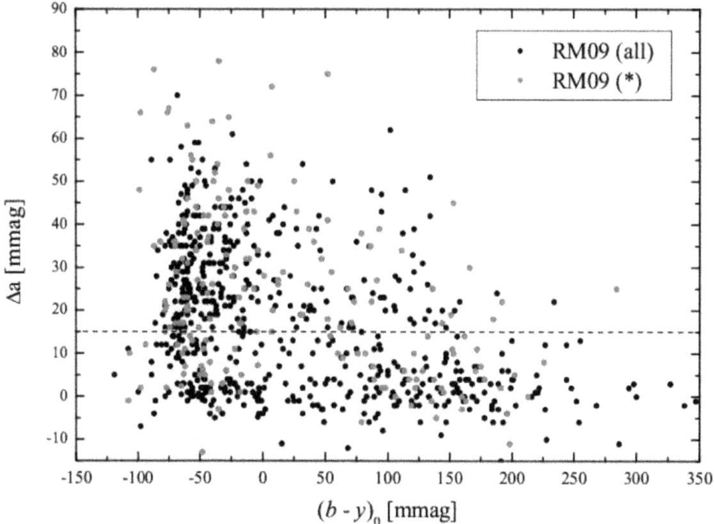

Figure 49: The Δa versus $(b-y)_0$ diagram for all stars included in Renson & Manfroid (2009, RM09, black circles). The stars marked therein as well-established are denoted with red circles. The dotted line marks the detection limit of CP stars according to Section 11.1.

an undetected bias because of an inadequate treatment. If this hypothesis is rejected, one can conclude that all well established CP stars in this colour range were successfully detected by the Δa photometric system. The apparent majority of CP stars with $(b-y)_0 < -0.05$ mag can not be argued by astrophysical means but is an manifestation of the intrinsic distribution of the observed stars (Fig. 22). Looking for the most extreme Δa values larger than +50 mmag, they are found for stars with $(b-y)_0 < +0.13$ mag (spectral type of about A7). The majority of these objects are B-type stars which is line with the predictions of the detailed stellar atmospheric models (Sect. 14.9.2). A detailed analysis of the detection probabilities of the different CP groups is given in Section 11.

Magnetic field strength: the catalogue by Bychkov et al. (2009) was search for all entries of the complete sample of stars as listed in Section 10. Bychkov et al. (2009) published quadratic effective magnetic field $\langle B_e \rangle$ on the basis of different methods. For the definition of $\langle B_e \rangle$ and its correlation with the magnetic field modulus, for example, the reader is referred to Bychkov et al. (2003). If more than one mean value is available for a star, the following measurement (the flag is according to the used meth-

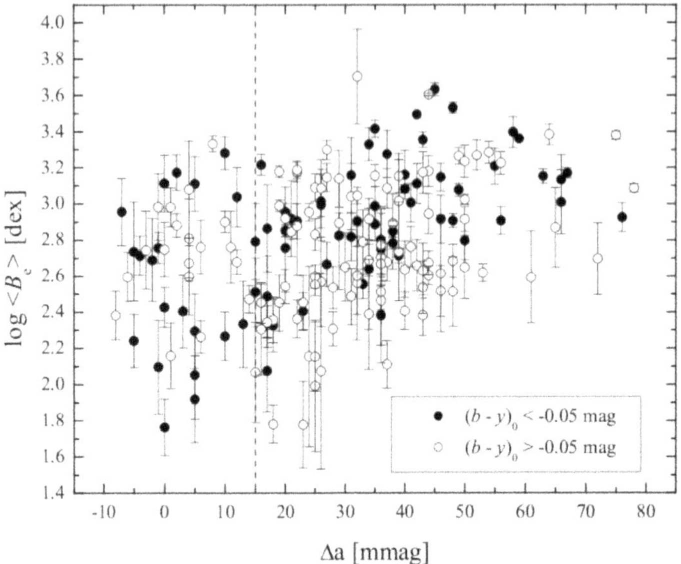

Figure 50: The $\log\langle B_{\rm e}\rangle$ (Bychkov et al., 2009) versus Δa diagram for the 187 stars with a measured Δa value (Sect. 10) and a magnetic field detection with $\sigma > 3$. The sample was divided according to a $(b-y)_0$ value of -0.05 mag which corresponds roughly to a spectral type of B9. The dotted line marks the detection limit of CP stars according to Section 11.1.

ods/lines) was used, sorted after the priority "all" ⇒ "met" ⇒ "Hl". This is a pure heuristic sequence based on the number of available measurements, Bychkov et al. (2003, 2009) list neither weights nor a discussion of the quality for the corresponding methods, for example. In total, 288 objects are included in both catalogues. From these, only 187 stars have a σ (the significance of the found magnetic field) larger than three. Figure 50 shows the results for the latter sample divided according to a $(b-y)_0$ value of -0.05 mag which corresponds roughly to a spectral type of B9. There is a clear correlation between $\log\langle B_{\rm e}\rangle$ and Δa independent of the colour evident. This result is well in line with the predictions from stellar model atmospheres (Sect. 14.9.2). From this figure it can be concluded that for all objects with a Δa value larger than +30 mmag, a $\langle B_{\rm e}\rangle$ larger than 200 G was unambiguously detected. This lower limit for the magnetic field is statistically highly significant because there are 105 stars with $\Delta a > +30$ mmag in this sample and range. From these, 94 have a measured $\langle B_{\rm e}\rangle$ larger than 200 G. From the remaining 11 stars, only one object

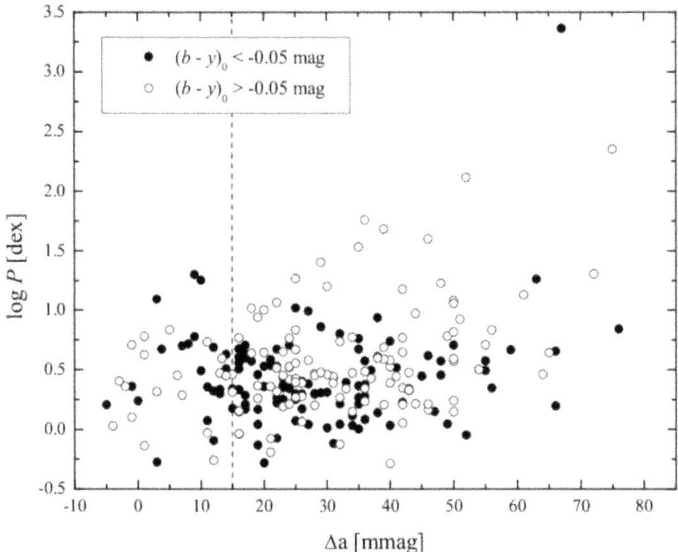

Figure 51: The $\log P$ versus Δa diagram for 238 stars with a measured Δa value (Sect. 10) and a published rotational period from VSX. The sample was divided according to a $(b-y)_0$ value of -0.05 mag which corresponds roughly to a spectral type of B9. The dotted line marks the detection limit of CP stars according to Section 11.1.

(HD 74169) has a significant lower $\langle B_e \rangle$ with 130(40) G, five objects (HD 6164, HD 44456, HD 91239, HD 137193, and HD 161277) have insignificant magnetic field measurements lower than even 100 G, but five objects (HD 37140, HD 136347, HD 150035, HD 164429, and HD 219749) have mean values between 250 and 950 G, respectively. Note that there is, as far as it is known, no contamination of supergiants (Sect. 11.8) at such large Δa values. Only Pleione reached during its shell phase, once, an exceptional Δa value of +36 mmag (Sect. 11.9). Therefore, the contamination of objects other than CP ones in this spectral type range is almost zero. This shows the uniqueness of the Δa photometric system to trace stellar magnetic fields for stars of the upper MS with a statistically high probability.

Rotational period: Photometric variability of the CP2 star α^2 Canum Venaticorum (ACV) was first reported by Guthnik & Prager (1914). The light curves of CP stars can be fitted well by a sine wave and its first harmonic with varying amplitudes depending on the photometric filter systems

(Bernhard et al., 2015a,b). For some CP stars, a double-wave structure of the photometric light curves depending on the observed wavelength region is detected (Maitzen, 1980b). However, similar magnetic field modulus variations are rather rare exceptions (Mathys et al., 1997). The variability of CP stars is explained in terms of the oblique rotator model (Stibbs, 1950), according to which the period of the observed variations is simply the rotational period. Accurate knowledge of the rotational period and its evolution in time is a fundamental step in understanding the complex behaviour of CP2/4 stars, especially as far as it concerns the phase relation between the magnetic, spectral, and light variations (Mikulášek et al., 2010). The "AAVSO International Variable Star Index"; (VSX, Watson, 2006) database was consulted to check for entries for all stars included in RM09 with measured Δa values as variable. From this list, all stars not classified as ACV or ROT variables were deleted. These were mainly pulsating, δ Scuti, γ Doradus, and Slowly Pulsating B-type, stars but also some elliptical variables as well as eclipsing binaries. The final sample includes 238 stars. Already Kodaira (1973) speculated if there is an anti-correlation between the rotational period and the degree of the chemical peculiarity. In the light of the theory of magnetic accretion (Havnes, 1974) this could be understood by the loss of angular momentum to the local environment and thus enhanced accretion which leads to a stronger chemical peculiarity. Therefore, slower rotating CP stars should have stronger pronounced peculiarities. In the light of the diffusion theory (Alecian, 2015), slower rotation means less influence of the meridional circulation (mixing) and a further stabilisation of the magnetic field which makes the diffusion more efficient. This effect is called magnetic breaking. Netopil (2013) presented a comprehensive overview of the literature and made a thoroughly analysis of the available data. He concluded that there seems to be a correlation between the investigated parameters: equatorial velocity, magnetic field strength, and iron abundance. But these properties were also influenced by variations with the phase of the rotation period. Furthermore, there were strong indications that the magnetic field strength declines with increasing equatorial velocity, or vice versa the stronger the magnetic field, the slower the stellar rotation. This would support the theory of magnetic braking, which occurs probably already during the pre-MS phase. For the T_{eff} range between 9000 and 12 000 K there were indeed indications that slower rotations implies higher iron abundances (one of the main contributers to the Δa index). But the effects were not very pronounced. Figure 51 shows the $\log P$ versus Δa diagram for the 238 stars with a measured Δa value (Sect. 10) and a published rotational period from VSX. The sample was divided according to a $(b-y)_0$ value of -0.05 mag which corresponds roughly to a spectral type of B9. It has to be emphasized that such a subdivision is only a first approximation to stellar evolutionary effects. Although it is statistically significant, no clear trend can be concluded from this sample. There seems no obvious strong correlation between the rotational period and the strength of the Δa index which is not in favour of magnetic braking.

16 Investigating the λ Bootis spectroscopic binary hypothesis using Δa photometry

Here, a spin-off of the Δa photometric system and its measurements is presented. This shows its capability not only to detect different non-normal star groups but also to investigate their characteristics in more detail.

More than 60 years after the first notification of the peculiar nature of λ Bootis (HR 5351) by Morgan et al. (1943), the origin and even the existence of a homogeneous group of λ Bootis stars (Sect. 6) is still a matter of debate.

The issue still remains whether the group itself is homogeneously defined as addressed by Gerbaldi et al. (2003). Their working group has formulated the hypothesis that the tendency to detect lower abundances for numerous elements in high resolution spectroscopy abundance analysis might be explained by assuming an unidentified (or unidentifiable) binary system of two "normal" (solar abundant) stars with similar spectral type. As a consequence they conclude that the group of λ Bootis stars consists of single type objects and an unknown, but probable high percentage of undetected spectroscopic binary systems.

In the following this hypothesis was investigate by comparing synthetic photometric indices of binary systems to those of apparent group members.

16.1 Modelling the spectroscopic binary systems

The working hypothesis is rather simple: a spectroscopic binary system with two solar abundant components has to simulate the total energy flux distributions and thus the photometric colours of an apparent single λ Bootis star.

To investigate the chances of λ Bootis stars being disguised binary systems, we modelled a set of stars spanning the range in fundamental parameters (Table 22) populated by λ Bootis stars (Paunzen et al., 2002a).

The turbulent velocity was fixed at a value of 2 km s^{-1}, convection was modelled according to Canuto & Mazzitelli (1991, hereafter CM) since

Table 22: Parameter space used for the models and the astrophysical parameters of the standard star Vega and the λ Bootis stars HD 107233 and HD 204041.

	Models	Vega	107233	204041
T_{eff} [K]	7000 – 9000	9560	6900	8100
$\log g$ [cgs]	3.50 – 4.50	4.05	3.80	4.10
v_{turb} [km s^{-1}]	2	2	3	2
Convection	Canuto & Mazzitelli (1991)			

Table 23: Location of the stars according to Figure 52. A plus sign means that the objects are located within the area of the spectroscopic binary models. The results for the individual stars are listed in Table 24.

Diagram	(1)	(2)	(3)	(4)	(5)	(6)
Δa	–	–	–			
$\Delta(V1-G)/Z$	–	–	+/–	+/–	+	
m_1/m_2	–		–	–	+	
m_1						+/–

the stars are situated in the region of weak to no convection. To model the atmospheres as well as the energy flux distribution of these stars and of the photometric anchor Vega, the code LLmodels v.SE/8.0 (Shulyak et al., 2004) was used. This model atmosphere code supports individual chemical composition, VCS theory (Vidal et al., 1973; Lemke, 1997) for the treatment of Hydrogen lines and the CM convection treatment. To represent the atomic line opacities a separate line list for each model was created. The line parameters where obtained form the Vienna Atomic Line Database (Kupka et al., 1999). All lines for which $l_\nu/\alpha_\nu \geq 1\%$ (l_ν and α_ν are line and continuum absorption coefficient) was realized for the given model structure were selected. The step size in wavelength of the synthetic spectra was set to 0.1 Å.

An energy flux distributions of 105 hypothetical binary stars was built by preparing all possible combinations of always two spectra. On these, synthetic photometry for the Strömgren $uvby\beta$, Δa, and the Geneva 7-colour system was performed. As it is common practice, Vega was the reference point for the synthetic photometry. This stars atmosphere was modelled according to the parameters of Hill & Landstreet (1993), the observed photometric indices were inquired from the SIMBAD/CDS and

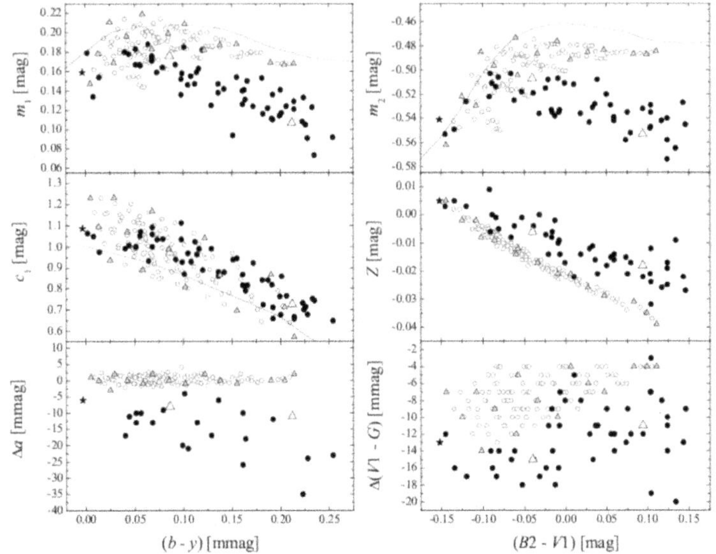

Figure 52: A comparison of the synthetic Strömgren $uvby\beta$, Δa (left panels) and the Geneva 7-colour system (right panels) with the observations for the group of λ Bootis stars (full circles). The open circles are the binary models whereas the grey shaded triangles are the single star models used to construct them. Vega is marked as asterisk, the two synthetic "standard" λ Bootis stars are open triangles. The standard lines are from the literature.

GCPD astronomical databases.

To show also a comparison to synthesized intrinsic λ Bootis stars, the stars HD 107233 and HD 204041 were modelled with the parameters and abundances listed in Heiter et al. (2002a).

For the statistical analysis, no reddening was taken into account. Gerbaldi et al. (2003) have extensively discussed the problems using the standard Strömgren $uvby\beta$ dereddening procedures for the group of λ Bootis stars. Their values were compared with those of Paunzen et al. (2002a) and a good agreement was found. However, the reddening for most objects is negligible because they are all located in the solar neighbourhood.

16.2 Results and Conclusions

Figure 52 shows the results of the relevant photometric diagrams. The observational data for the λ Bootis stars were taken from Paunzen et al. (2002a). The classical metallicity sensitive diagrams m_1 versus $(b - y)$ and m_2 versus $(B2 - V1)$ show that for cooler objects, the λ Bootis stars are nicely separated from the models on the MS, for example c_1 versus $(b - y)$. These observational facts were also described in Paunzen et al. (1997).

Very interesting are the other three diagrams which include directly metallicity depend indices (Sect. 4). The most important index is Δa, for which, unfortunately, the least measurements are available whereas $\Delta(V1 - G)$ seems less sensitive with a larger scatter. Notice that Vega as metal-deficient object (Hill & Landstreet, 1993; García-Gil et al., 2005) shows, as expected, a reduced Δa value. Otherwise, the standard lines and observations of Vega fit the synthetic values very good. The differences of the synthetic and observed values for the two λ Bootis stars are only 1 to 2 mmag for Δa, Z, and $\Delta(V1 - G)$, respectively.

According to Fig. 52, the sample was divided in six different groups numbered from (1) to (6) depending if the placement is compatible with the spectroscopic binary models taking an error of ± 5 mmag into account.

The final division is listed in Table 23 whereas Table 24 shows the results for the individual stars. The first two groups include objects which are not located in the spectroscopic binary area (SBA hereafter) in any diagram. The third and fourth group include stars which are situated in the SBA either in the $\Delta(V1 - G)$ or Z diagram, but not in the classical metallicity and the Δa diagram or do not have an available Δa measurement (fourth group).

The fifth group comprise good candidates for undetected spectroscopic binaries among the λ Bootis group because these stars compare well to the synthetic photometry of the binary models. HD 198160 is already known as close visual binary star whereas inconsistent $v \sin i$ measurements were reported by Heiter (2002) for HD 170680. The four objects of this group certainly deserve further attention in the future.

The last group includes those stars which could only be tested within the m_1 versus $(b - y)$ diagram which prevents any clear conclusion.

If one compares the list of Table 24 with the results of Faraggiana et al. (2004), only two stars, HD 11413 and HD 210111, were reported as spectroscopic binary systems on the basis of radial velocity shifts of two and three spectra, respectively. Both are well investigated δ Scuti pulsators (Breger et al., 2006) which makes them especially interesting for further investigations. HD 210111 comprises of two equal (within the estimated

Table 24: The results for λ Bootis stars taken from Paunzen et al. (2002a) according to Fig. 52 and Table 23. Only the stars of group (5) can be considered as being good candidates for undetected spectroscopic binary systems. No conclusion can be drawn for group (6), the abbreviations in (7) are **P**assed or **F**ailed. The last row contains the number of objects as part of the total sample.

(1)	(2)	(3)	(4)	(5)	(6)	(7)
6870	319	31295	111604	23392	13755	**P**
11413	7908	35242	156954	110377	15165	**P**
24472	30422	101108	193256	170680	54272	**P**
75654	74873		193281	198160	87271	**P**
81290	91130				90821	**F**
83041	110411				105759	**P**
83277	125162				111005	**P**
84123	130767				120500	**F**
102541	183324				120896	**P**
105058	204041				175445	**F**
106223	221756					
107233						
109738						
125889						
142703						
142994						
149130						
153747						
154153						
168740						
168947						
184779						
192640						
210111						
216847						
44 %	19 %	5 %	7 %	7 %	18 %	

errors) underabundant λ Bootis-type stars (Paunzen et al., 2012). This object is very similar to the systems HD 84948 and HD 171948 investigated by Iliev et al. (2002).

The percentage of undetected spectroscopic binary systems mimicking a single, metal-weak object seems very low. From 47 well investigated stars, groups (1) to (5), only four objects seem good candidates for a further investigation which is below 10% of the complete sample.

A carefully preselection of λ Bootis stars, using Δa photometry results in a homogeneous group of intrinsic λ Bootis stars which can be used to investigate the group properties in more detail.

17 The Δa observations of Galactic open clusters

Since the 1970ies, CP stars in open clusters were targets of, first, photoelectric (Maitzen, 1981) and then CCD (Bayer et al., 2000) observations. Up to now, fifteen parts of photoelectric (last: Paunzen et al., 2014) and eight parts of CCD (last: Netopil et al., 2007) Δa photometry of Galactic open clusters are published by now (Table 25). In addition, there is one paper published by Netopil et al. (2005) which presented CCD Δa and BVR photometry of NGC 7296.

Netopil (2013) performed a deep and comprehensive study of the a-priori detected CP stars and their characteristics in these open clusters. The interested reader is referred to this work.

Here, the focus is set on how to conduct photoelectric and CCD Δa observations as well as how to interpret the different diagrams. Still, there are a lot of unpublished photoelectric and CCD measurements in the archive available which await for reductions and publication.

In addition, the list of already published aggregates with their parameters are given for both samples. First, a short overview about the advantages of observations of open clusters are listed.

17.1 The characteristics of open clusters

Open clusters are physically related groups of stars held together by mutual gravitational attraction. Therefore, these populate a limited region of space, typically much smaller than their distance from the Sun, so that the members are all approximately at the same distance (Lata et al., 2002). They are believed to originate from large cosmic gas and dust clouds (diffuse nebulae), and to continue to orbit the host galaxy through the disk.

Table 25: The list of the fifteen papers including photoelectric (upper panel) and the eight with CCD (lower panel) Δa observations of Galactic open clusters.

I	Maitzen & Hensberge (1981)	IX	Maitzen & Pavlovski (1987a)
II	Maitzen & Floquet (1981)	X	Jenkner & Maitzen (1987)
III	Maitzen (1982)	XI	Maitzen & Schneider (1987)
IV	Maitzen & Wood (1983)	XII	Maitzen & Pavlovski (1987b)
V	Maitzen & Schneider (1984)	XIII	Maitzen et al. (1988)
VI	Maitzen (1985)	XIV	Maitzen (1993)
VII	Maitzen et al. (1986)	XV	Paunzen et al. (2014)
VIII	Maitzen & Catalano (1986)		
I	Bayer et al. (2000)	V	Paunzen et al. (2003a)
II	Paunzen & Maitzen (2001)	VI	Paunzen et al. (2005a)
III	Paunzen & Maitzen (2002)	VII	Paunzen et al. (2006)
IV	Paunzen et al. (2002b)	VIII	Netopil et al. (2007)

In many clouds visible as bright diffuse nebulae, star formation still takes place, so that one can observe the birth of new young star clusters (Massi et al., 2015). This process of formation takes only short time (a few Myrs) compared to the lifetime of the cluster, so that, speaking in terms of stellar evolution, all member stars are of similar age. Also, as all the stars in a cluster are formed from the same diffuse nebula, they are all of similar initial chemical composition (Netopil et al., 2016). Hence, open clusters are of great interest for several fields of astrophysics because:

- The cluster members are all at about the same distance from the Sun.
- Within a few million years cluster members have approximately the same age: a small range compared to the age of all clusters older than a billion years.
- The chemical composition of the members is quite homogeneous within a cluster, but the metallicities of the various open clusters range from about -1.0 to $+0.6$ dex compared to the Sun.
- The member stars have different masses, ranging from about 120 $M_\odot$ for the most massive stars in very young clusters to less than about 0.08 $M_\odot$

Distance, age and metallicity are, in general, not straightforward to determine for galactic field stars. Open clusters, on the other hand, represent samples of stars of constant age and homogeneous chemical composition, suited for the study of processes linked to stellar structure and evolution, and to fix lines or loci in several most important astrophysical diagrams such as the colour-magnitude diagram (CMD), or the Hertzsprung-Russell diagram (HRD). Comparing the "standard" HRD, derived from nearby stars with sufficiently well known distances, or the theory of stellar evolution, with the measured CMD of star clusters, provides a considerably good method to determine the distance of star clusters. Comparing their HRD with stellar theory provides the best way to estimate the age of star clusters. The fact that all the cluster HRDs can be explained by the theory of stellar evolution gives convincing evidence for our theories and models, and moreover for the underlying physics including nuclear and atomic physics, quantum physics and thermodynamics. In those cases where certain detailed features of cluster HRDs deviate from theoretical predictions, this gives an incentive to review and improve the parts of the stellar evolution models which cause the deviations. Observations of open clusters thus enables to enhance the knowledge about the physics of stellar evolution.

About 2800 open clusters are known in the Milky Way (Kharchenko et al., 2013), and this is probably only a small percentage of the total population. Most of them are much too distant and therefore too faint to be observable, or have only a short life as stellar swarms. As they drift along their orbits, some of their members escape the hosting cluster due to velocity changes in mutual closer encounters, tidal forces in the galactic gravitational field and encounters with field stars and interstellar clouds crossing their way (Chumak et al., 2010). A typical open cluster has lost most of its member stars along its path after several hundred Myrs. Only some of them reach an age older than a few billion years. The escaped individual stars continue to orbit the Milky Way on their own as field stars. It is suspected that most or even all field stars in our and the external galaxies have their origin in clusters.

17.2 Photoelectric Δa observations of Galactic open clusters

In a series of fifteen papers (Table 25), photoelectric Δa observations of Galactic open clusters were published.

Based on the example of four open clusters (Feinstein 1, NGC 2168, NGC 2323, and NGC 2437), the typical reduction process and detailed

Table 26: Fundamental parameters of the target clusters taken from Paunzen & Netopil (2006) and Zejda et al. (2012).

Cluster		$\alpha(2000)$	$\delta(2000)$	$R_\odot$ [pc]	$E(B-V)$ [mag]	*age* [Myr]	$\mu_\alpha \cos\delta$ [mas/yr]	μ_δ [mas/yr]
Feinstein 1	C1103−595	11 05 56	−59 49 00	1180	0.41	5	−6.1	+2.9
NGC 2168	C0605+243	06 09 00	+24 21 00	830	0.23	100	+1.5	−2.9
NGC 2323	C0700−082	07 02 42	−08 23 00	895	0.23	100	+0.4	−2.0
NGC 2437	C0739−147	07 41 46	−14 48 36	1495	0.16	235	−5.0	+0.4

Table 27: Observations log, the description of the used equipment can be found in the last column.

Cluster	Observatory	Telescope	Time	Reference
Feinstein 1	ESO	Bochum 0.61 m	89/04	Maitzen (1993)
NGC 2168	Hvar	0.65 m	89/01, 89/02	Maitzen & Pavlovski (1987b)
NGC 2323	ESO	0.5 m	85/02, 85/03	Maitzen & Schneider (1987)
NGC 2437	ESO	1.0 m	84/02	Maitzen (1993)

analysis is presented.

17.2.1 Observations and reduction

This work was originally initiated by the European working group on CP stars of the upper main sequence (Mathys et al., 1989). The regular photoelectric observations of this programme were performed until 1991. Nevertheless, there are no doubts about the accuracy of the photoelectric measurements.

The fundamental parameters of the target clusters taken from Paunzen & Netopil (2006) and Zejda et al. (2012) are listed in Table 26. The observations log is listed in Table 27.

Standard stars were usually measured 10 to 20 s per filter, programme stars of 120 s per filter. In several cases the latter were observed only 20 to 30 s per filter in the corresponding filter sequences which were repeated several times. This way inhomogeneities of the sky were thought to be averaged out more effectively. The raw data were corrected for dead-time depending on the used photomultiplier tube, converted to magnitudes, and corrected for night-to-night variations and extinction, and averaged over the individual observations.

To deredden the programme stars, use was made of photometric data in the Johnson, Geneva, and Strömgren systems, compiled from the open cluster database WEBDA[2]. For the first two systems, the well-known calibrations based on the X/Y parameters (Cramer, 1999) and Q-index (Gutierrez-Moreno, 1975), respectively, applicable for O/B-type stars were applied. Objects with available Strömgren data were treated with the routines by Napiwotzki et al. (1993), allowing the dereddening of cooler type stars. For member stars without sufficient photometry, the mean cluster reddening was adopted. There are only a few non-member stars (fifteen in total), for which it was not possible to derive individual reddening values; these were therefore rejected in the subsequent analysis. If several estimates for a particular object are available, a mean value was calculated. Since all clusters are rather near to the Sun, a strong reddening especially for cool-type stars is hardly expected.

Because of the reddening, the normality line is shifted by $E(B-V)$ to the red and by a small amount $E(a)$ to higher a-values (Sect. 3). The mean ratio of these shifts $f = E(a)/E(B-V)$ was determined as $f \approx 0.035$. The final values were calculated as $a_{\mathrm{corr}} = a(\mathrm{obs}) - fE(B-V)$. For the subsequent analysis, the dereddened diagnostic a_{corr} versus $(B-V)_0$ diagrams (Fig. 53) were used. The mean $(B-V)$ values have been calculated from the data included in WEBDA.

The normality line for each cluster was determined using the photometric data of member and non-member stars, which is justified because of the use of individually dereddened colours. Objects deviating more than 3σ were rejected in an iterative process. The final coefficients of $a_{\mathrm{corr}} = b + c(B-V)_0$ together with their errors are listed in Table 28. An object is considered as positively (or negatively) detected if its Δa value, taking into account the observational error, lies above (or below) the 3σ limit of the corresponding normality line. In the following, the numbering system from WEBDA (W no.) is used.

For the membership probabilities of the individual stars, the method given in Balaguer-Núnez et al. (1998) was employed. In addition, the results were compared with those of the algorithm published by Javakhishvili et al. (2006), yielding excellent agreement. The first method takes both the errors of the mean cluster and the stellar proper motions into account. The mean proper motions of the target clusters were taken from Zejda et al. (2012) and are listed in Table 26. The proper motions of the individual stars were taken from the following sources, sorted by the priority:

- TYCHO-2 (Høg et al., 2000);
- UCAC4 (Zacharias et al., 2013);

[2]http://webda.physics.muni.cz

Table 28: Final results, all photometric values are given in mmag. For Feinstein 1 W17 it was not possible to derive reliable astrophysical parameters (see text). The errors in the final digits of the corresponding quantity are given in parenthesis.

	Feinstein 1	NGC 2168	NGC 2323	NGC 2437
$a_{corr} = b + c(B-V)_0, N$	17(2)/55(6)/24	29(2)/47(17)/21	−34(4)/35(17)/17	123(1)/61(6)/37
3σ	±8	±9	±14	±12
$N(>50\%)/N(<50\%)$	22/6	19/8	9/9	36/12
#CP/Δa/Prob	W16/+16/99	W244/+16/93	W171/+51/82	W172/+34/80
	W17/+15/59	W336/+17/37		W457/+26/76
		W364/+23/69		W469/+38/63
				W476/+19/50
$\log T_{eff}/\log L/L_{\odot}$	4.350/3.36	4.126/2.32	4.105/2.07	3.993/1.67
	−/−	4.060/2.25		4.018/1.82
		4.086/2.77		3.978/1.98
				3.985/1.72

- PPMXL (Röser et al., 2010).

For the complete sample, kinematic data are available. A comparison with the results published by Baumgardt et al. (2000) yields an excellent agreement.

The effective temperature calibrations and bolometric corrections for CP stars by Netopil et al. (2008) were applied on the available photometric data in the Johnson, Geneva, and Strömgren systems. The luminosities were derived using the cluster parameters listed in Table 26 regardless of the kinematic non-membership in order to obtain an additional membership criterion. In general, an uncertainty for the derived temperatures of 500 K and 700 K for CP2 and CP4 stars were adopted, respectively (Netopil et al., 2008). The errors in luminosity were derived using a standard error in distance of 10 %, 0.1 mag for bolometric correction, and 0.02 mag for brightness and interstellar reddening.

By comparing the derived absolute magnitudes and temperatures with available spectral types, it was concluded that all kinematic non-members can be also considered cluster non-members from the photometric point-of-view.

One exceptional case is Feinstein 1 W17 which is a kinematical member, but from photometry it seems the contrary. A detailed analysis for it is given in Section 17.2.2.

From the final a_{corr} versus $(B-V)_0$ diagrams (Fig. 53), ten stars were found that deviate significantly from the corresponding normality lines. Only two of them seem not to be a member of their associated star cluster.

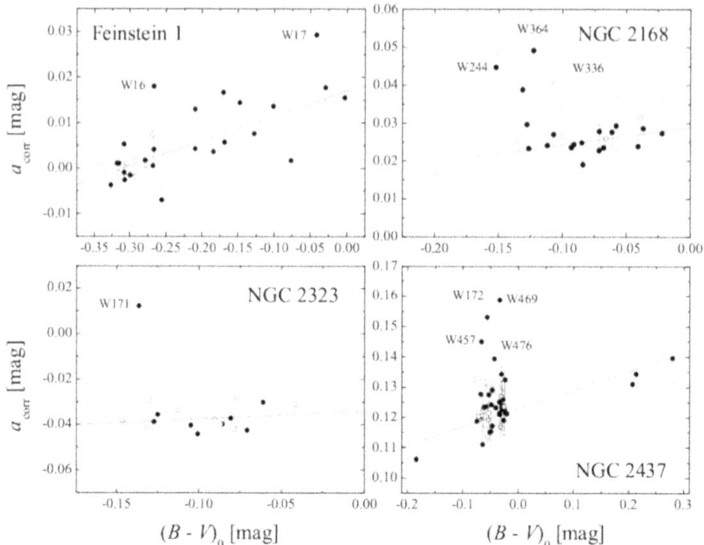

Figure 53: a_{corr} versus $(B-V)_0$ diagrams for the target clusters listed in Table 26. Filled circles denote stars with a kinematical membership probability of more than 50%, open circles less than 50%. Stars with a statistical significant Δa value, are denoted with their WEBDA numbers (W no.). The solid line is the normality line.

However, since each star was dereddened individually, these stars are still very good candidates for being true CP stars. In the following, the results for the open clusters are discussed in more detail.

17.2.2 Feinstein 1

This aggregate is an accumulation of brighter stars around the β Cephei variable and helium star HD 96446 (W20). If this is indeed a true star cluster, it is very young and all A- and F-type stars should still be in their Pre-MS phase (Feinstein, 1964). Since its initial discovery, it has hardly been studied.

Two stars, HD 305941 (W16) and HD 306034 (W17), were found significantly above the normality line. Garcia (1993) classified W16 as B2 IV/V ($v \sin i$ of $80\,\mathrm{km\,s^{-1}}$) and W17 as Am (kA2 IV, mF6 III/IV). The projected rotational velocity of W16 is exceptionally low for an early B-type star which already points to a CP nature. They also noted that "He II at

4009Å presents intensity variations". From the classification of W17 it is already clear that this is a very extreme CP1 star.

From the kinematic data, the membership probabilities of 99% and 59% were derived, respectively. Therefore, these stars were calibrated using the reddening and distance of Feinstein 1 (Table 26). This resulted in $T_{\rm eff}$ and $M_{\rm Bol}$ of [22 400 K, −3.7 mag] and [10 300 K, −1.6 mag] for W16 and W17. While the values for W16 are perfectly in line with the spectral type, they are not compatible for W17 even if a pre-main-sequence status is assumed. For W17, a different approach on the basis of Johnson UBV photometry was chosen. Taking the standard relation for $(B-V)$ versus $(U-B)$ from Schmidt-Kaler (1982) and the observed values, the reddening was estimated assuming that the object follows the standard relation. As a result, $E(B-V)=0.15$ mag was calculated. With this reddening value, the $T_{\rm eff}$ was calibrated as 7800 K. This value is well in line with the spectral type of Am (kA2 IV, mF6 III/IV). As a last step, the $T_{\rm eff}$ value was kept fixed and the distances were calculated which place the star either at the zero- or terminal-age-main-sequence within the evolutionary grids by Schaller et al. (1992). The derived values are between 300 pc and 850 pc which establishes it as being a foreground star. Recalculating the Δa value using the reddening given above and the relation from Table 28 yields +10 mmag. However, this peculiar object is worthwhile for follow-up spectroscopic observations. One can conclude that W17 is not associated with Feinstein 1, although its kinematic characteristic is, to a certain degree, compatible with the mean cluster proper motion.

From its stellar parameters, it can be inferred that W16 is a CP4 (He-strong) star with about 7.7 $M_\odot$. This star is the most massive and most luminous star among the CP candidates. Cidale et al. (2007) presented a similar diagram for a sample of CP4 stars (see Fig. 7 therein). From this diagram, it was found that this star is among the youngest CP4 objects known so far.

17.2.3 NGC 2168

For this open cluster, two CP star candidates are listed in the literature (Niedzielski & Muciek, 1988): W364 (HD 252405) and W547 (HD 252459). The source of the corresponding information is unknown. Both stars are classified in Hoag & Applequist (1965) as B6 III: and A3 V, respectively. The star W364 has a measured Δa value of +23 mmag which confirms its peculiar nature. For W547, +17 mmag is measured, but the error of the mean is too large to make it statistically significant. One could argue that this is caused by an apparent photometric variability due to spots and rotation. However, according to the kinematical data, W547 is not a member (1% probability) of NGC 2168. On the other hand, W364 is a member of the cluster. It is close to the Terminal-Age-MS with a mass of about 4.1 $M_\odot$. This would correspond to a probable B7 Si-type star.

In addition, two other stars from our sample are significantly above the normality line, namely W244 (TYC 1877−356−1, +16) and W336 (HD 252427, +17). For neither of these objects spectral classifications are available (Skiff, 2016). The kinematical data supports the membership of W244 (93%), but not for W336 (37%). They both have masses between 3 $M_\odot$ and 4 $M_\odot$ and would be therefore classified as late B-type Si stars. Note that HD 252427 is incorrectly identified as W335 in SIMBAD/CDS. For W336, the star HD 252458 is listed.

The only known emission-type object in the cluster area (Kohoutek & Wehmeyer, 1999), HD 41995 (W781), was not measured.

17.2.4 NGC 2323

The known CP candidate stars within NGC 2323 are HD 52965 (W3) and BD−08 1708 (W51). Bychkov et al. (2009) listed upper limits of the magnetic fields for these objects of 89 G and 49 G, respectively. They list a spectral classification of B8 Si, taken from Renson & Manfroid (2009), for HD 52965 which is probably wrong. The only found reference in this respect is from Young & Martin (1973) who list “B9 p: Si II (4128/4130) slightly enhanced”. A Δa value of +3 mmag was measured. This value together with the non-detection of a magnetic field suggests that this object is not a CP star. Young & Martin (1973) published the spectral type of “B9p:: Hg II 3984Å weak” for BD−08 1708. A classification which is not typical for a CP3 object. However, this star was not measured.

For the star CD−08 1704 (W171) a significant positive Δa value of +51 mmag was found. Its membership probability is 82%. The are some investigations of NGC 2323 that include photometric but no spectroscopic data (Clariá et al., 1998; Sharma et al., 2006; Frolov et al., 2012). Clariá et al. (1998) list $E(B-V) = 0.23$ mag for W171 which corresponds to the mean value of the cluster (Table 26). The star W171 is a 100 Myr old, 3.4 $M_\odot$ (late B-type), probable Si, star.

17.2.5 NGC 2437

The open cluster NGC 2437 has been studied photometrically several times because it might host a planetary nebula (Majaess et al., 2007). However, there is no detailed spectroscopic analysis available in the literature. Neither CP nor Be stars within the cluster area have been detected so far.

Four stars were detected that lie above the normality line, namely W172 (TYC 5422−967−1, $\Delta a = +34$ mmag), W457 (TYC 5422−2127−1, +26), W469 (BD−14 2133, +38), and W476 (TYC 5422−305−1, +19). For none of these objects are spectral classifications available in the literature. All stars are, within the errors, well represented by the isochrone for NGC 2437 with a distance of 1500 pc and an age of 235 Myr. The derived masses range from 2.4 $M_\odot$ to 2.9 $M_\odot$ which corresponds to early A-types.

17.3 CCD Δa observations of Galactic open clusters

The large photoelectric survey in the Δa system was mainly dedicated to clusters no more distant than 1000 pc from the Sun (Sect. 17.5). Nevertheless, these results still serve as an important basis for new CCD observations in order to provide photometric standards, for example. With the CCD technique, it is possible to observe open clusters much more distant than that providing a more general representation of the Galactic field.

Nowadays, one can hardly find any photomultiplier tubes (PMT) in operation at international observatories. The CCD technique has developed very rapidly over the last twenty years. But even early comparisons between PMTs and CCDs were in favour of the latter, both in accuracy and efficiency (Abbott & Kleinman, 1994).

However, the CCD technology not only has advantages but also has pitfalls. Here a short list of things are listed, the user of CCD Δa observations has to keep in mind.

- Because the normality line is deduced from the apparent normal type stars of the same frame where the apparent peculiar objects are located, photometry can also be performed in non-photometric conditions. The only assumption is that the absorption due to clouds, for example, is homogeneous over the whole frame and the exposure time (Croom et al., 1999).

- The observation of bright and faint stars with the same exposure time is very critical. Normally, each CCD chip has a limited linearity range. Therefore, for a given exposure time, bright stars in the field are saturated and faint ones are underexposed. To overcome this problem, one has to use at least two different exposure times and then merge the frames accordingly.

- Normally, the field of view of modern instruments with CCD is limited to about 25', except of CCD arrays. This means that if one wants to observe near and large clusters, the mosaicing technique has to be applied. It is a well known technique but care has to be taken that enough common stars on the fields are available or in other words, there is a sufficient overlap.

In a series of eight papers (Table 25), CCD Δa observations of Galactic open clusters were published. In addition, there is one paper published by Netopil et al. (2005) which presents CCD Δa and BVR photometry for NGC 7296.

The reduction of the individual photometric CCD frames of the different filters follow the common ways of such a framework. There are many books, recipes, and guides available in the literature (Howell, 2006). Here only the very basic and common steps are mentioned (see also Sect. 18.1).

- Subtract dark from all images, but bias frames
- Combine all bias files
- Subtract bias from all images
- Combine flats for a mean flat field
- Apply mean flat field to all scientific images
- Get photometric values using aperture and/or point spread function fitting for all objects
- Search for possible zero point offsets and/or trends between the photometric values of the individual frames and correct for them
- Get mean photometric values for all objects
- Generate the different photometric plots

The analysis of the observational material is in general identical as listed in Section 17.2. In Fig. 54, three typical examples of diagnostic diagrams within the Δa photometric system for three open clusters (NGC 6134, NGC 6192, and NGC 6451) are shown. Their cluster parameters are listed in Table 30. From this figure several conclusions can be drawn which are very important when working with the Δa photometric system. The individual cluster diagrams are described in more detail in the following.

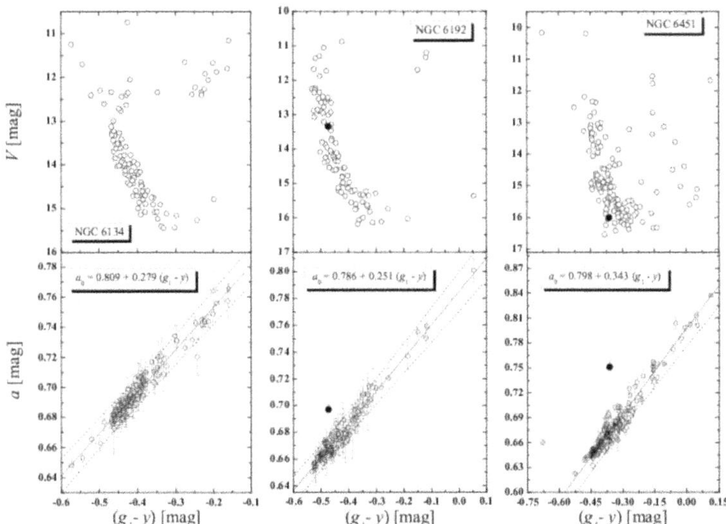

Figure 54: Typical diagnostic diagrams within the Δa photometric system. Here, three open clusters (NGC 6134, NGC 6192, and NGC 6451) observed with the CCD technology are shown. The solid line is the normality line whereas the dotted lines are the confidence intervals corresponding to 99.9 %.

- NGC 6134: There is a well defined MS, turn-off point, and the giant population for this cluster. Several Blue Stragglers (BS, Ahumada & Lapasset, 2007) and one very bright non-member are visible. Neither of these stars exhibit a significant deviating a value which is line with the investigation listed in Section 11.8.

- NGC 6192: Again, the MS and three giants are clearly visible, but no BS are present. One CP star was detected in this cluster.

- NGC 6451: Here, a typical, more general case for distant open cluster is shown. There are two bright non-members which are probable foreground stars. Notice that the bluer of these objects show a significant positive Δa value. But with the help of the CMD, this object can be unambiguously identified as non-member and not as cluster CP star candidate. In addition, several more reddened background stars are present. None of them show a deviating a value because this diagram is almost insensitive to reddening (Sect. 3).

These examples illustrate the variety of possible characteristics of the diagnostic diagrams. However, normally, the true nature (membership, fore- and background as well as bona-fide CP stars) of the deviating cases can be solved with the help of these diagrams, alone without the help of additional information such as astrometric or spectroscopic data.

17.4 The observation of Blue Stragglers in open clusters

As described in Sect. 7.1, Blue Stragglers (BS) were discovered as members of open and globular clusters which are hotter (bluer) than the turn-off point of the corresponding cluster (Mermilliod, 1982).

Pendl & Seggewiss (1975) were among the first who investigated the spectra of these stars in more detail. Moreover, they were testing the hypothesis if BS are CP stars which would explain their bluer colour. This "blueing" effect is typical for some magnetic CP objects due to stronger UV absorption than in normal type stars (Adelman, 1980). They investigated BS which are true members of IC 4756. They found a percentage of up to 58% of CP stars among BS. Later on, Abt (1985) found even a higher incidence of 62%. These numbers are much higher than the normal incidence of CP stars among all MS objects (a maximum of 15%) of the same spectral type (Sect. 6). However, these findings were somehow contradicted by a series of papers by Schönberner et al. (2001) who derived detailed elemental abundances for BS of open clusters. Although they found several He peculiar objects, the iron and titanium abundances agreed with those of the normal cluster members. The projected rotational velocities of the confirmed BS were significantly lower than the mean values typical for their spectral types. This would give a hint that diffusion might play a significant role in their atmospheres. Up to now, no detectable (strong) magnetic fields of BS have been found yet (Mathys, 1988).

Maitzen et al. (1981) studied 27 BS within nine open clusters via photoelectric Δa measurements in order to derive the CP incidence among this sample. This was, until now, the only dedicated investigation of this star group.

Their target clusters were all older than 100 Myr which corresponds to a BS population of a spectral type between B3 and F0. The targets were selected on the basis of their location in the HRD as well as radial velocity and proper motion data to guarantee a high membership probability. They found two stars (8.3% of the total sample) with strong and another four (17%) with a mild photometric variability on the basis of the dereddened

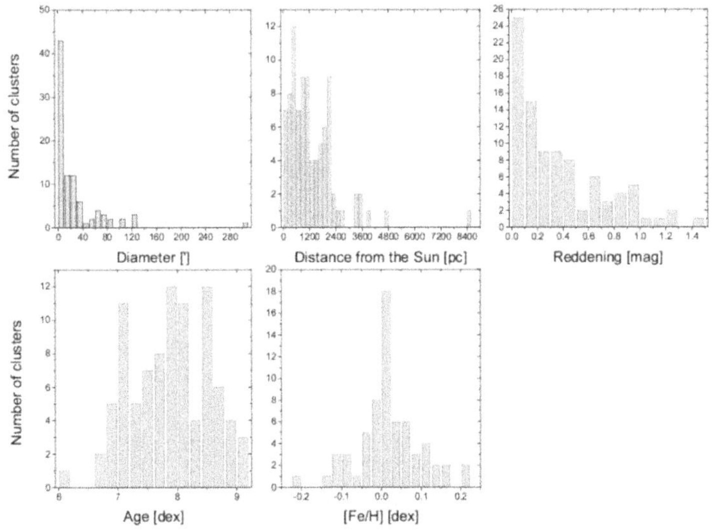

Figure 55: The histograms of the cluster parameters taken from Table 30.

normality line (Sect. 3). This number is very much in line with the "normal overall" incidence of CP stars (see discussion above). But one has to keep in mind that this conclusion is based on somehow poor number statistics.

17.5 Statistics of the observed open clusters

In this section, an overview of all, up-to-now, observed Galactic open clusters is given. This includes all photoelectric and CCD observations. The cluster parameters were taken from DAML02, version 3.5 (Dias et al., 2002) whereas the metallicity ([Fe/H]) is from Netopil et al. (2016). The Cartesian $[XYZ]$ coordinates are defined as

- X: radially outwards away from Galactic Centre
- Y: in direction of Sun's rotation around the Milky Way
- Z: out of Galactic plane, positive towards North Galactic Pole

All values are listed in Tables 29 and 30. The histograms of the cluster parameters are shown in Figure 55. From this figure, the following conclusions can be drawn

Table 29: All observed aggregates with the references listed in Table 25.

Cluster	α (2000) [h]	δ (2000) [°]	l [°]	b [°]	X [pc]	Y [pc]	Z [pc]	peo	CCD
Berkeley 11	04 20 36	+44 55 00	157.0815	−03.6421	−2022	+855	−140		VII
Berkeley 94	22 22 42	+55 51 00	103.0946	−01.1844	−596	+2561	−54		VII
Collinder 121	06 56 20	−24 43 46	235.7016	−09.9906	−610	−895	−191	IV	
Collinder 132	07 15 20	−30 41 00	242.9893	−08.8024	−212	−416	−72	XIV	
Collinder 140	07 24 27	−31 51 00	244.9298	−07.5912	−170	−364	−54	X	
Collinder 272	13 30 26	−61 19 00	307.5947	+01.2016	+1278	−1660	+44		IV
Feinstein 1	11 05 56	−59 49 00	290.0251	+00.3847	+397	−1089	+8	XV	
Haffner 15	07 45 32	−32 51 00	247.9516	−04.1576	−1310	−3235	−254		VII
IC 2391	08 40 32	−53 02 00	270.3622	−06.8387	+1	−174	−21	VIII	
IC 2602	10 42 58	−64 24 00	289.6014	−04.9061	+54	−151	−14	XIII	
IC 4665	17 46 18	+05 43 00	30.6191	+17.0820	+290	+171	+103	V	
IC 4725	18 31 47	−19 07 00	13.7023	−04.4335	+601	+146	−48	VI	
King 21	23 49 54	+62 43 00	115.9465	+00.6827	−920	+1891	+25		VIII
Lynga 1	14 00 02	−62 09 00	310.8494	−00.3373	+1243	−1437	−11		VII
Lynga 14	16 55 04	−45 14 00	340.9189	−01.0885	+832	−288	−17		IV
Melotte 20	03 24 19	+49 51 42	146.5676	−05.8625	−154	+101	−19	XII	
Melotte 22	03 47 00	+24 07 00	166.5707	−23.5211	−119	+28	−53	IX	
Melotte 105	11 19 42	−63 29 00	292.9039	−02.4119	+667	−1578	−72		I
Melotte 111	12 25 06	+26 06 00	221.3538	+84.0248	−8	−7	+95	IX	
NGC 1039	02 42 05	+42 45 42	143.6580	−15.6132	−387	+285	−134	VII	
NGC 1502	04 07 50	+62 19 54	143.6724	+07.6577	−798	+587	+133		VI
NGC 1662	04 48 27	+10 56 12	187.6950	−21.1142	−404	−55	−157	I, IV	
NGC 1901	05 18 11	−68 27 00	279.0397	−33.6065	+60	−378	−255	XIV	
NGC 2099	05 52 18	+32 33 12	177.6354	+03.0915	−1380	+57	+75		V
NGC 2168	06 08 54	+24 20 00	186.5910	+02.1913	−905	−105	+35	XV	
NGC 2169	06 08 24	+13 57 54	195.6076	−02.9345	−1012	−283	−54	XIV	I
NGC 2232	06 27 15	−04 45 30	214.4324	−07.5422	−294	−201	−47	X	
NGC 2287	06 46 01	−20 45 24	231.0196	−10.4446	−439	−543	−129	IV	
NGC 2323	07 02 42	−08 23 00	221.6722	−01.3311	−709	−631	−22	XV	
NGC 2343	07 08 06	−10 37 00	224.2681	−01.1728	−756	−737	−22	X, XIV	
NGC 2362	07 18 41	−24 57 18	238.1789	−05.5482	−777	−1252	−143	III	
NGC 2422	07 36 35	−14 29 00	230.9582	+03.1303	−308	−380	+27	IV	
NGC 2423	07 37 06	−13 52 18	230.4836	+03.5369	−486	−590	+47	XIV	
NGC 2437	07 41 46	−14 48 36	231.8576	+04.0644	−930	−1185	+107	XV	
NGC 2439	07 40 45	−31 41 36	246.4425	−04.4666	−518	−1188	−101		III
NGC 2447	07 44 30	−23 51 24	240.0387	+00.1346	−518	−898	+2	XIV	
NGC 2451A	07 43 12	−38 24 00	252.5754	−07.2986	−56	−179	−24	VIII	
NGC 2451B	07 44 27	−37 40 00	252.0509	−06.7264	−92	−285	−35	VIII	
NGC 2489	07 56 15	−30 03 48	246.7128	−00.7731	−1564	−3634	−53		II
NGC 2516	07 58 04	−60 45 12	273.8158	−15.8558	+26	−393	−112	I, IV	
NGC 2546	08 12 15	−37 35 42	254.8552	−01.9859	−240	−887	−32	III	
NGC 2547	08 10 09	−49 12 54	264.4649	−08.5975	−34	−355	−54	XV	
NGC 2567	08 18 32	−30 38 24	249.7951	+02.9610	−578	−1572	+87		II
NGC 2632	08 40 24	+19 40 00	205.9196	+32.4843	−142	−69	+100	XII	
NGC 2658	08 43 27	−32 39 30	254.5554	+06.0698	−535	−1937	+214		II
NGC 3105	10 00 39	−54 47 18	279.9147	+00.2636	+1469	−8403	+39		VI
NGC 3114	10 02 36	−60 07 12	283.3317	−03.8395	+210	−884	−61	XIII	V
NGC 3228	10 21 22	−51 43 42	280.6971	+04.5453	+101	−533	+43	III	
NGC 3293	10 35 51	−58 13 48	285.8562	+00.0736	+636	−2238	+3		VIII

Table 29: continued.

Cluster	α (2000) [h]	δ (2000) [°]	l [°]	b [°]	X [pc]	Y [pc]	Z [pc]	peo	CCD
NGC 3532	11 05 39	−58 45 12	289.5708	+01.3468	+165	−463	+12	XI	
NGC 3960	11 50 33	−55 40 24	294.3672	+06.1826	+759	−1675	+199		III
NGC 4103	12 06 40	−61 15 00	297.5736	+01.1551	+755	−1446	+33	XV	
NGC 5281	13 46 35	−62 55 00	309.1610	−00.7137	+700	−859	−14		II
NGC 5460	14 07 27	−48 20 36	315.7462	+12.6281	+489	−477	+153	VI	
NGC 5662	14 35 37	−56 37 06	316.9370	+03.3939	+486	−454	+39	XI	
NGC 5999	15 52 08	−56 28 24	326.0082	−01.9249	+1699	−1145	−69		VIII
NGC 6025	16 03 17	−60 25 54	324.5508	−05.8840	+613	−436	−78	XV	
NGC 6031	16 07 35	−54 00 54	329.2813	−01.4981	+1567	−931	−48		VII
NGC 6087	16 18 50	−57 56 06	327.7258	−05.4256	+750	−474	−84	VI	
NGC 6134	16 27 46	−49 09 06	334.9170	−00.1976	+1141	−534	−4		III
NGC 6192	16 40 23	−43 22 00	340.6473	+02.1223	+1459	−512	+57		III
NGC 6204	16 46 09	−47 01 00	338.5597	−01.0395	+1117	−439	−22		V
NGC 6208	16 49 28	−53 43 42	333.7545	−05.7648	+838	−413	−94		II
NGC 6250	16 57 56	−45 56 12	340.6832	−01.9196	+816	−286	−29		I
NGC 6268	17 02 10	−39 43 42	346.0486	+01.2998	+1048	−260	+24		VI
NGC 6281	17 04 41	−37 59 06	347.7306	+01.9725	+468	−102	+16	V	
NGC 6396	17 37 36	−35 01 36	353.9299	−01.7737	+1185	−126	−37		IV
NGC 6405	17 40 20	−32 15 12	356.5802	−00.7766	+486	−29	−7	V	VII
NGC 6451	17 50 41	−30 12 36	359.4775	−01.6005	+2079	−19	−58		III
NGC 6475	17 53 51	−34 47 36	355.8606	−04.5005	+299	−22	−24	II	
NGC 6611	18 18 48	−13 48 24	16.9540	+00.7935	+1722	+525	+25		IV
NGC 6633	18 27 15	+06 30 30	36.0112	+08.3278	+301	+219	+54	XV	
NGC 6705	18 51 05	−06 16 12	27.3070	−02.7759	+1666	+860	−91		V
NGC 6756	19 08 42	+04 42 18	39.0884	−01.6813	+1169	+950	−44		V
NGC 6802	19 30 35	+20 15 42	55.3256	+00.9167	+1011	+1462	+28		VIII
NGC 6830	19 50 59	+23 06 00	60.1353	−01.7997	+816	+1421	−51		VIII
NGC 6834	19 52 12	+29 24 30	65.6981	+01.1890	+850	+1883	+43		VII
NGC 7092	21 31 48	+48 26 00	92.4027	−02.2418	−14	+325	−13	VII	
NGC 7235	22 12 25	+57 16 12	102.7011	+00.7822	−732	+3248	+45		VI
NGC 7243	22 15 08	+49 53 54	98.8568	−05.5237	−124	+795	−78	XII	
NGC 7296	22 28 01	+52 19 22	101.8831	−04.5880	−503	+2390	−196		
NGC 7510	23 11 03	+60 34 12	110.9026	+00.0645	−1242	+3251	+4		VI
Pismis 20	15 15 23	−59 04 00	320.5164	−01.2002	+2525	−2080	−69		IV
Ruprecht 44	07 58 51	−28 35 00	245.7456	+00.4823	−1943	−4312	+40		VIII
Ruprecht 115	16 12 52	−52 24 00	330.9588	−00.8493	+1888	−1048	−32		VIII
Ruprecht 120	16 35 10	−48 17 00	336.3854	−00.4932	+1832	−801	−17		VIII
Ruprecht 130	17 47 32	−30 06 00	359.2214	−00.9600	+2100	−29	−35		VII
Stock 2	02 14 43	+59 29 06	133.3339	−01.6942	−208	+220	−9	XV	
Stock 16	13 19 29	−62 38 00	306.1486	+00.0631	+1068	−1462	+2		VI
Trumpler 2	02 36 53	+55 54 54	137.3765	−03.9710	−532	+490	−50	XV	
Trumpler 10	08 47 54	−42 27 00	262.7907	+00.6740	−53	−421	+5	X	

Table 30: The cluster parameters of all observed aggregates (Table 29). The apparent diameter (d), distance from the Sun ($R_{\odot}$), colour excess $E(B-V)$ in $(B-V)$, and age ($\log t$) were taken from DAML02, version 3.5 (Dias et al., 2002) whereas the metallicity ([Fe/H]) is from Netopil et al. (2016).

Cluster	d	$R_{\odot}$	$E(B-V)$	$\log t$	[Fe/H]	Cluster	d	$R_{\odot}$	$E(B-V)$	$\log t$	[Fe/H]
	[']	[pc]	[mag]	[dex]	[dex]		[']	[pc]	[mag]	[dex]	[dex]
Berkeley 11	5	2200	0.950	8.041	−0.01	NGC 3114	35	911	0.069	8.093	+0.05
Berkeley 94	3	2630	0.608	6.996		NGC 3228	5	544	0.028	7.932	+0.01
Collinder 121	60	1100	0.040	7.080		NGC 3293	6	2327	0.263	7.014	
Collinder 132	80	472	0.037	7.080		NGC 3532	50	492	0.028	8.477	0
Collinder 140	60	405	0.030	7.548	+0.01	NGC 3960	5	1850	0.290	9.100	−0.04
Collinder 272	10	2095	0.604	7.020	+0.03	NGC 4103	6	1632	0.294	7.393	
Feinstein 1	25	1159	0.400	7.000		NGC 5281	7	1108	0.225	7.146	−0.02
Haffner 15	4	3500	1.050	7.300		NGC 5460	35	700	0.092	8.200	+0.03
IC 2391	60	175	0.008	7.661	−0.01	NGC 5662	29	666	0.311	7.968	+0.04
IC 2602	100	161	0.024	7.507	−0.02	NGC 5999	3	2050	0.450	8.600	−0.03
IC 4665	70	352	0.174	7.634	−0.03	NGC 6025	14	756	0.159	7.889	
IC 4725	29	620	0.476	7.965	0	NGC 6031	3	1823	0.371	8.069	0
King 21	4	2103	0.886	7.163	+0.03	NGC 6087	14	891	0.175	7.976	+0.21
Lynga 1	3	1900	0.450	8.000	+0.02	NGC 6134	6	1260	0.350	8.950	+0.11
Lynga 14	3	881	1.428	6.712		NGC 6192	9	1547	0.637	8.130	+0.12
Melotte 20	300	185	0.090	7.854	+0.14	NGC 6204	5	1200	0.460	7.900	+0.02
Melotte 22	120	133	0.030	8.131	−0.01	NGC 6208	18	939	0.210	9.069	−0.01
Melotte 105	5	1715	0.830	8.550	+0.06	NGC 6250	10	865	0.350	7.415	
Melotte 111	120	96	0.013	8.652	0	NGC 6268	6	1080	0.400	7.600	+0.14
NGC 1039	35	499	0.070	8.249	+0.02	NGC 6281	8	479	0.148	8.497	+0.06
NGC 1502	8	1000	0.700	7.000		NGC 6396	3	1192	0.926	7.506	
NGC 1662	20	437	0.304	8.625	−0.11	NGC 6405	20	487	0.144	7.974	+0.07
NGC 1901	10	460	0.030	8.780	−0.08	NGC 6451	7	2080	0.672	8.134	+0.01
NGC 2099	14	1383	0.302	8.540	+0.02	NGC 6475	80	301	0.103	8.475	+0.02
NGC 2168	40	912	0.200	8.250	−0.21	NGC 6611	6	1800	0.800	6.110	
NGC 2169	5	1052	0.199	7.067		NGC 6633	20	376	0.182	8.629	−0.08
NGC 2232	53	359	0.030	7.727	+0.11	NGC 6705	32	1877	0.428	8.400	+0.12
NGC 2287	39	710	0.010	8.400	−0.11	NGC 6756	4	1507	1.180	7.790	+0.08
NGC 2323	14	950	0.200	8.000		NGC 6802	5	1778	0.840	8.980	+0.01
NGC 2343	5	1056	0.118	7.104	+0.03	NGC 6830	5	1639	0.501	7.572	+0.22
NGC 2362	5	1480	0.100	6.700		NGC 6834	5	2067	0.708	7.883	
NGC 2422	25	490	0.070	7.861	+0.09	NGC 7092	29	326	0.013	8.445	0
NGC 2423	12	766	0.097	8.867	+0.08	NGC 7235	5	3330	0.900	6.900	
NGC 2437	20	1510	0.100	8.400	−0.07	NGC 7243	29	808	0.220	8.058	+0.03
NGC 2439	9	1300	0.370	7.000		NGC 7296	8	2450	0.240	8.450	−0.02
NGC 2447	10	1037	0.046	8.588	−0.05	NGC 7510	6	3480	0.900	7.350	
NGC 2451 A	108	302	0.055	7.648	0	Pismis 20	4	3272	1.280	6.864	
NGC 2451 B	120	189	0.010	7.780	−0.08	Ruprecht 44	10	4730	0.619	6.941	
NGC 2489	6	3957	0.374	7.264	+0.05	Ruprecht 115	5	2160	0.650	8.780	+0.01
NGC 2516	30	409	0.101	8.052	+0.05	Ruprecht 120	3	2000	0.700	8.180	
NGC 2546	70	919	0.134	7.874	+0.01	Ruprecht 130	3	2100	1.200	7.700	0
NGC 2547	25	361	0.186	7.585	−0.14	Stock 2	60	303	0.380	8.230	+0.17
NGC 2567	7	1677	0.128	8.469	−0.04	Stock 16	3	1810	0.520	6.900	
NGC 2632	70	187	0.009	8.863	+0.16	Trumpler 2	17	725	0.400	7.950	
NGC 2658	6	2021	0.043	9.152	−0.01	Trumpler 10	29	424	0.034	7.542	−0.12
NGC 3105	2	8530	0.950	7.300							

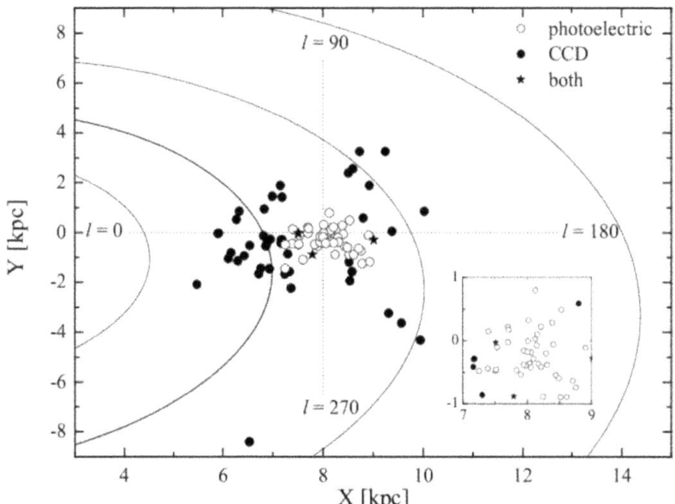

Figure 56: Galactic location of the published open clusters observed via Δa photometry projected to the Galactic plane. The Galactic centre is located at $(X/Y) = (0/0)$ kpc and the Sun at $(X/Y) = (8/0)$ kpc. Galactic longitudes in degrees are indicated for better orientation. Furthermore, the spiral arm model by Bobylev & Bajkova (2014) is shown as solid lines. The spiral arms from the inner to the outer area are Scutum-Crux, Carina-Sagittarius, Perseus, and Cygnus.

- d: because the apparent diameters of open clusters are directly connected to the distance (the absolute diameters do not exceed 20 pc), the peak at less than 10' is due to the clusters observed via the CCD technology.

- $R_\odot$: most clusters are located within 2 kpc around the Sun. Only a few are more distant than that. Clearly, more distant clusters are needed to be observed to complete a statistically significant sample.

- $E(B-V)$: clusters with reddening values of up to 1.5 mag were observed and also CP stars have been detected. This shows the capability of the Δa to function even within such highly reddened environments.

- $\log t$: clearly, there is a lack of very young (younger than 10 Myr) investigated clusters. This is mainly because there are almost no close and young clusters in the solar neighbourhood (photoelectric survey) existing. With the CCD technology, already several very young aggregates have been observed.

- [Fe/H]: only clusters with ± 0.2 dex have been observed so far. For further observations, the metallicity range has to be significantly increased.

Figure 56 shows the Galactic location of the sample projected to the Galactic plane together with the spiral arm model published by Bobylev & Bajkova (2014). The spiral arms from the inner to the outer area are Scutum-Crux, Carina-Sagittarius, Perseus, and Cygnus. The regions with the highest reddening (Neckel et al., 1980) around l = [90,270] are still not investigated in more detail because the lack of observable star clusters. However, the solar neighbourhood around 1 kpc is already well analysed. One also have to keep in mind the survey of Galactic field stars (Sect. 10) within this region.

18 The first Δa observations of three Galactic globular clusters

Globular clusters are main astrophysical laboratories to test and modify evolutionary models. They are extensively used to place constraints on key ingredients of canonical stellar evolution models, such as the mixing length parameter of convective energy transport theory (Ferraro et al., 2006). They are also rather "simple" stellar systems consisting of a distinct population, which is in dynamic equilibrium. Therefore, they are used for extensive N-body simulation in order to understand the formation and evolution of the Milky Way (Sippel et al., 2012). There are about 160 Globular clusters known in the Milky Way (Harris, 1996, 2010 edition).

It is well known that several globular clusters have at least two main sequences (MS), which are explained by a different helium content (Piotto et al., 2007). In addition, different red giant (RGB) and sub-giant branches (SGB) were also found (Piotto et al., 2012). A similar characteristic of the horizontal-branch (HB), namely at least two different populations, was detected by Grundahl et al. (1998). There are different "jumps" in the blue HB (BHB) distribution in the V versus $(u - y)$ colour-magnitude diagram (CMD), which could be used to select apparent peculiar objects (Grundahl et al., 1999). Also, peculiar HB extensions, like the blue-hook were found (Brown et al., 2001). However, the cause of these phenomena is still not

Table 31: The basic cluster parameters of the targets taken from Harris (1996, 2010 edition).

NGC	Name	$\alpha(2000)$ [°]	$\delta(2000)$ [°]	l [°]	b [°]	$R_{\odot}$ [kpc]	R_{GC} [kpc]	$E(B-V)$ [mag]	[Fe/H] [dex]
104	47 Tuc	00 24 05.67	−72 04 52.6	305.89	−44.89	4.5	7.4	0.04	−0.72
6205	M13	16 41 41.24	+36 27 35.5	59.01	+40.91	7.1	8.4	0.02	−1.53
7099	M30	21 40 22.12	−23 10 47.5	27.18	−46.84	8.1	7.1	0.03	−2.27

Table 32: Observing log for the programme clusters. The clusters were observed by I.Kh. Iliev (II) and O.I. Pintado (OP).

Cluster	Site	Date	Obs.	$\#_{g_1}$	$\#_{g_2}$	$\#_y$	t_{g_1} [s]	t_{g_2} [s]	t_y [s]
NGC 104	CASLEO	08.2001	OP	6	10	10	5x60, 1x120	5x60, 5x120	5x120, 5x300
NGC 6205	BNAO	07.2003	II	12	12	13	12x100	12x100	13x100
NGC 7099	CASLEO	08.2001	OP	15	16	15	10x60, 6x120	10x60, 5x120	10x60, 5x180

clear, but it is probable connected to the complex star formation history of the individual clusters (Valcarce et al., 2012). In Sect. 7.5 an overview of the interesting stellar groups are given.

Here, as a pioneer study, the first photometric Δa observations of three, widely different, globular clusters NGC 104, NGC 6205, and NGC 7099 are presented. It is a vital test, if the Δa system is able to detect peculiar stars both at the HB and RGB, G- as well as K-type MS stars. While it is not expected to find traces of the different star sequences due to the fact that helium does not contribute in the 5200Å region at all.

In total, photometry of 2266 stars from 109 individual frames are presented. According to the 3σ detection limit of each globular cluster, 61 objects with positive and 29 with negative Δa values were detected. This corresponds to an upper limit of about 3% of apparent peculiar objects.

For NGC 6205, it was possible to compare the results with abundance determinations from the literature. The Δa values of three HB stars, one without chemical peculiarities, listed in Behr (2003a), are in perfect agreement. The peculiar objects were clearly detected with +57 and +60 mmag, whereas the non-peculiar object does not stand out. In addition, 10 RGB stars with [Mg/Fe] values from −0.15 to +0.30 dex, published by Johnson et al. (2005) were analysed. None of them exhibit a significant Δa value, which means that the elemental peculiarity of magnesium has to be much larger to be detected.

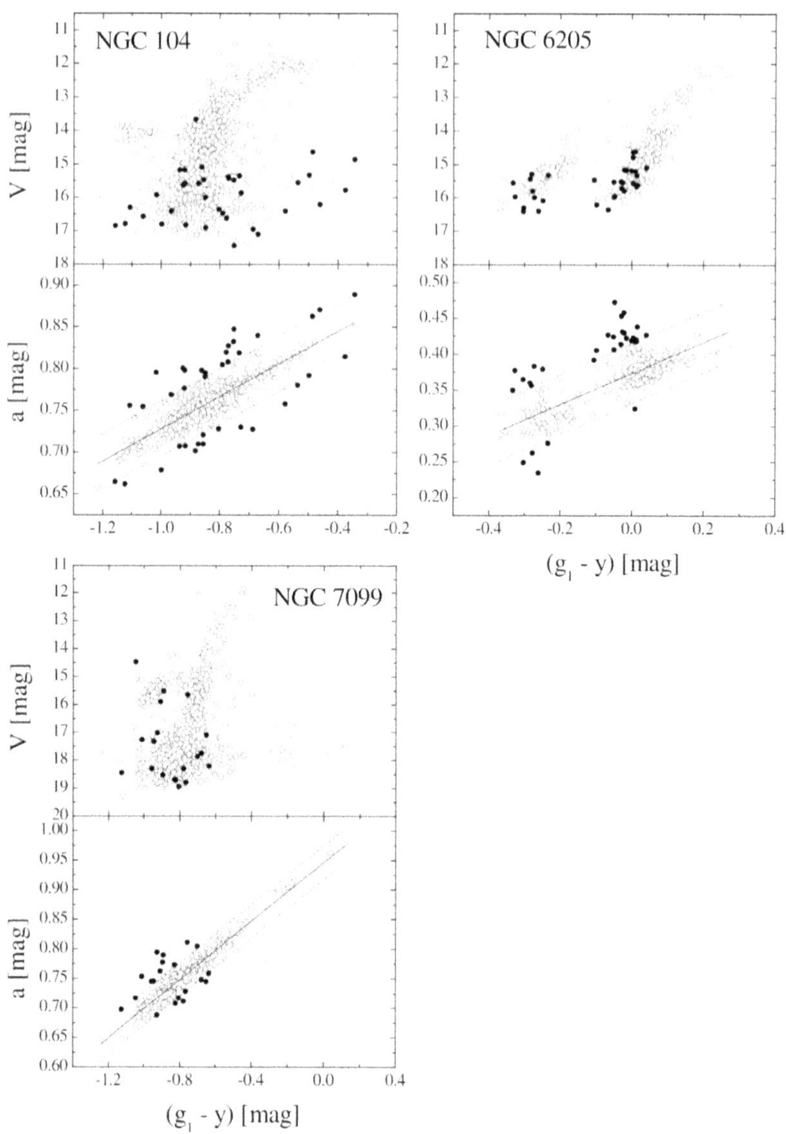

Figure 57: The photometric diagrams of the three globular clusters. The dotted lines are the 3σ upper and lower limits of the Δa detection sensitivity according to Table 33. The vertical lines are the locus of $(B-V)_0 = 1.0\,\mathrm{mag}$.

18.1 Target selection, observations and reduction

For this first case study, three globular clusters with widely different overall metallicities, namely NGC 104, NGC 6205, and NGC 7099, were selected. Such a criterion guarantees to detect possible peculiar stars for various local environments. Table 31 lists the basic cluster parameters of the targets, taken from Harris (1996, 2010 edition).

The observations of the three globular clusters were performed at two different sites:

- 2 m RCC telescope RCC telescope of the Bulgarian National Astronomical Observatory (BNAO, Rozhen), direct imaging, SITe SI003AB 1024 × 1024 pixel CCD, 5' field of view, 1 pixel = 0.32",
- 2.15 m telescope of the El Complejo Astronómico El Leoncito (CASLEO, San Juan), direct imaging with focal reducer, TEK-1024 CCD, 9.5' field of view, 1 pixel = 0.813".

The observing log with the number of frames in each filter and the integration times is listed in Table 32. The typical seeing conditions were between one and two arc seconds. The observations were performed with two different filter sets, both having the following characteristics: g_1 ($\lambda_c = 5007$ Å, FWHM = 126 Å, $T_P = 78\%$), g_2 (5199, 95, 68) and y (5466, 108, 70).

The CCD reductions were performed with standard IRAF v2.12.2a routines. All images were corrected for bias, dark, and flat-field. The photometry is based on point spread function fitting (PSF). For each image, at least 20 isolated stars were selected to calculate the individual PSFs. A Moffat15 function fitted the observations best. In the following, only stars were used which are detected on all frames. Because of instrumentally induced offsets and different air-masses between the single frames, photometric reduction of each frame was performed separately and the measurements were then averaged and weighted by their individual photometric error. The used photometric errors are based on the photon noise and the goodness of the PSF fit as described in Stetson (1987).

For NGC 6205 three different, overlapping, fields around the centre were observed. No significant photometric offsets between the fields were detected. The most inner parts (radius of about 1.5') of NGC 104 and NGC 7099 were not used for the further analysis because of the severe crowding and the unresolved single star content.

There is no significant reddening towards the targets (see Table 31). Assuming that all stars exhibit the same interstellar reddening and metallicity, peculiar objects deviate from the normality line more than 3σ (Sect. 3).

Table 33: Summary of results. The errors in the final digits of the corresponding quantity are given in parenthesis.

	NGC 104	NGC 6205	NGC 7099
$V = a + b \cdot (y)$	−3.23(21)/0.909(22)	−2.82(16)/0.965(9)	−5.10(19)/1.009(17)
Reference	Hesser et al. (1987)	Grundahl et al. (1998)	Alcaino et al. (1998)
$a_0 = a + b \cdot (g_1 - y)$	0.922(3)/0.194(4)	0.373(1)/0.217(9)	0.946(2)/0.248(3)
3σ [mag]	0.033	0.042	0.028
n(obj)	1107	365	794
n(positive)	21	28	12
n(negative)	16	5	8
n(frames)	26	37	46

For the transformation of the instrumental y to V magnitudes, the following references were used: Hesser et al. (1987, NGC 104), Grundahl et al. (1998, NGC 6205), and Alcaino et al. (1998, NGC 7099). Note that the zero-points for the measurements taken at BNAO and CASLEO are not the same due to different CCD gain and bias levels as well as extinction coefficients.

All results are summarized in Table 33. In total, 2266 stars on 109 frames are finally analysed. The complete photometric data together with the coordinates are available in electronic form.

18.2 Discussion

The tool of Δa photometry measures any flux/spectral abnormalities in the 5200Å region (Sect. 8). Employing the Δa photometric system on globular clusters aims primarily towards two widely different star groups (Sect. 7.5).

1. Photospheric chemically peculiar HB stars

2. Peculiar RGB, G- and K-type MS stars

The first group shows enhancements of iron peak elements of up to three times solar, or 2 dex compared to the mean metallicity, whereas the latter could be detected by peculiarities of Mg I lines as well as MgH features around 5200Å. In addition, cool type dwarfs and giants can be sorted out by the different equivalent widths of their Mg features. However, such a distinction can also be easily done in the classical CMD.

In Fig. 57, the results of the photometric observations are presented. For each globular cluster, the V versus $(g_1 - y)$ and a versus $(g_1 - y)$

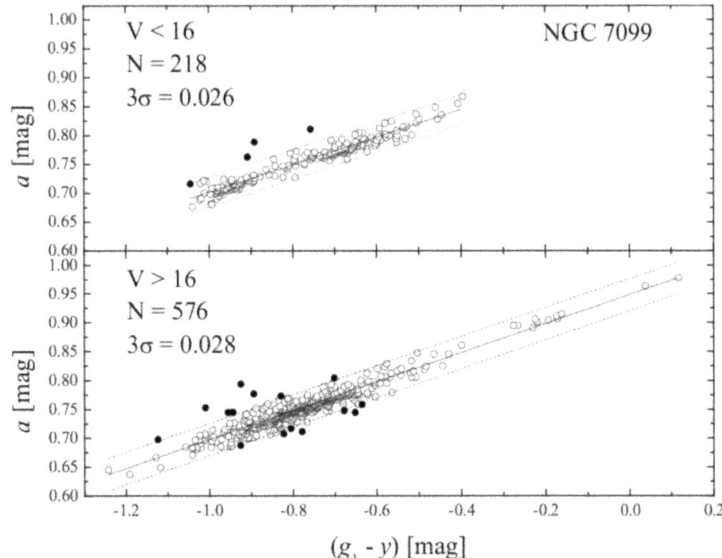

Figure 58: The a versus $(g_1 - y)$ diagram of NGC 7099. The sample was divided into stars which are brighter (upper panel) and fainter (lower panel) than 16th magnitude.

diagrams are shown. As expected, the CMDs, especially the characteristics of the HB, of the three aggregates are widely different.

The slopes of the normality lines range from 0.194 to 0.248 (Table 33), which is perfectly in line with values found for Galactic open clusters (Sect. 17). Notice that there seems to be a correlation with [Fe/H], i.e. NGC 104 with the highest metallicity exhibits the shallowest slope. However, with only three aggregates, one has to be careful with such a conclusion. Further observations are clearly needed to prove this apparent correlation.

Due to the photon noise, fainter stars have, in general, larger photometric errors. Inspecting the a versus $(g_1 - y)$ diagrams in Fig. 57, no correlation of the 3σ detection limit with the a values is visible. For NGC 7099, the detailed analysis is shown in Figure 58. The complete sample was divided in stars brighter and fainter than 16th magnitude. Both samples are rather different in terms of the number of stars and the range of $(g_1 - y)$. In comparison with the overall solution, only one object would not be detected as peculiar in the second sample with $\Delta a = +0.027$ mag which is just one mmag below the detection limit. In addition, the other samples were divided in different V subsamples and the 3σ detection limit

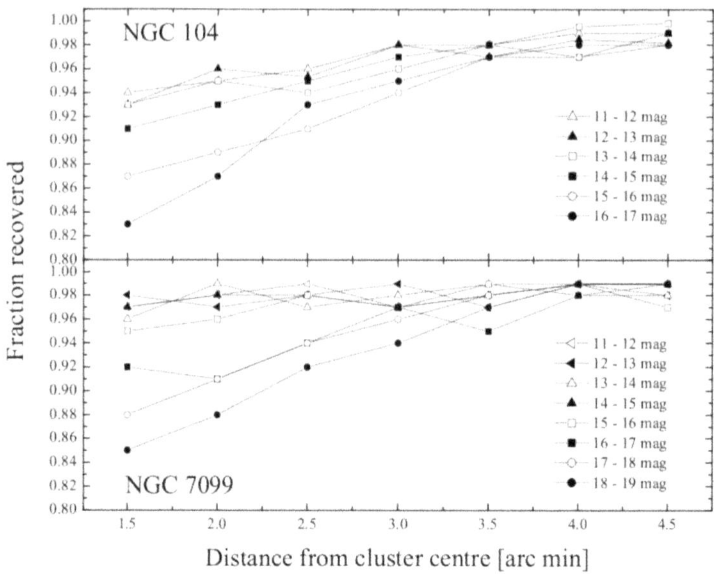

Figure 59: Results of the artificial-star tests for NGC 104 and NGC 7099. The inner most core parts with a radius of about 1.5' were not used for the analysis.

were calculated anew. All values agree within 1.5 mmag. Those tests justify using the sample as a whole for the analysis. This strong advantage of the Δa photometric system was already noticed before and is because for a given $(g_1 - y)$ value, a wide range of V values are sampled.

Also an artificial-star test to determine the completeness level of our sample was performed which is very important when analysing crowded field photometry (Anderson et al., 2008). Normally, faint stars in very crowded regions are either lost in the saturated cores or have be detected against the high background of these bright star aureoles. Thus, the magnitude limits for the detection of faint stars and the undercount correction estimates are functions of both the stellar magnitude and the distance of the objects from the most crowded and therefore bright cluster centre. As mentioned before, the inner-most core parts with a radius of about 1.5' were not used for the analysis. For this purpose, artificial stars were added with the IRAF task "addstar" to frames of each filter with the longest integration times. About 100 independent experiments, each consisting of 1000 artificial stars within a 1 mag interval randomly scattered throughout the image, were performed. These frames were then photometrically

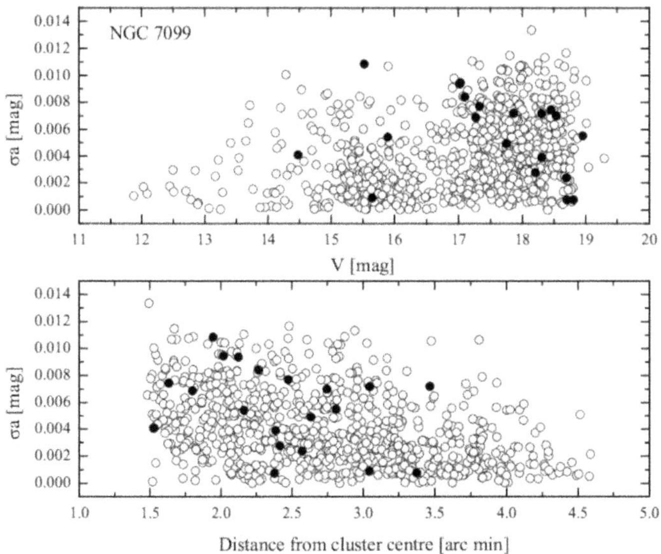

Figure 60: The differences of the a values from the artificial-star test and the observed ones (σa) versus the V magnitude (upper panel) and the distance from the centre (lower panel) for NGC 7099. The peculiar candidates (Fig. 57) are marked as filled circles.

analysed the same way as the original ones. The detected fraction of artificial stars was determined in concentric annuli of 0.5' width. Finally, a weighted average of the recovery fraction at each radius and magnitude interval was computed. For NGC 6205, we got almost a 99% completeness level for all bins because we observed fields quite far off the centre. Figure 59 shows the artificial-star tests for NGC 104 and NGC 7099. The band width of the individual curves is about ±3%. Since almost all regions and magnitude ranges are well above 90%, one can be confident that the effect of undetected binary stars do not play a significant role for the analysis. As a further test, the differences of the a values from the artificial-star test and the observed ones were calculated. Figure 60 shows these differences versus the V magnitude and the distance from the centre for the data of NGC 7099. Again, the distribution of the outliers does not significantly alter from the apparent normal type objects.

In total, photometric Δa values for 2266 stars were secured. According to the well established 3σ detection limit, 61 stars fall above, and 29 objects below the normality line. The latter can be also due to higher reddened

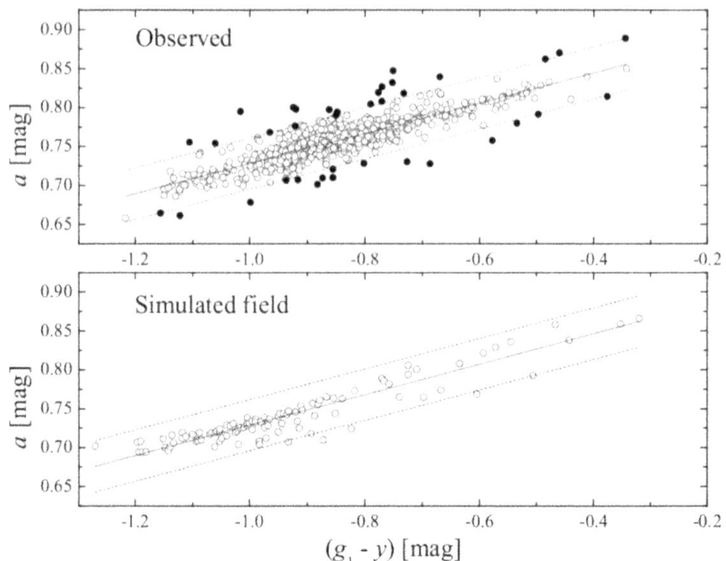

Figure 61: The simulated fore- and background population using the theoretical Galactic model, TRILEGAL (lower panel) and the observed one for NGC 104.

background stars, which would shift these objects below the normality line. The only reliable distinction could be done via membership probabilities based on proper motions. However, such kinematic data are not available for our target sample. Indeed, such field stars are present in globular cluster areas (Sariya et al., 2012). Those field stars are clearly visible in the right most upper panel of Fig. 5 in Sariya et al. (2012) where they exhibit redder $(V - I)$ colours than the cluster members. In the Δa photometric diagram, such stars could lie below the normality line, but not above it. The possibility of the influence of undetected visual companions for the outliers was also checked. In general, applying point-spread-function-fitting, already accounts for such cases. A comparison of "visual binaries" among normal type objects and outliers shows no significant accumulation among the latter.

To estimate the a versus $(g_1 - y)$ diagram of background/foreground stars in the field of view of our targets, the theoretical Galactic model, TRILEGAL 1.6[3] described by Girardi et al. (2005) was used. It includes the populations of the thin and thick disc as well as the Galactic halo

[3]http://stev.oapd.inaf.it/trilegal

but it is not able to simulate star clusters, yet. Fields of 12'x12' with the central coordinates as listed in Table 31 were simulated. The synthetic Δa photometry taken from the Vienna New Model Grid of Stellar Atmospheres, NEMO[4] (Heiter et al., 2002b) was used. The TRILEGAL output includes $T_{\rm eff}$, $\log g$, [Fe/H], and V for each object. First of all, the sample to the V magnitudes was restricted as deduced from Figure 57. Then, for each star, the closest model from the NEMO data base was searched for and the corresponding Δa photometry was taken. Figure 61 shows the observed and synthesized field of NGC 104. The situation for the other two fields is similar. The slope for the normality line of the observational data is 0.194(4) whereas it is 0.196(5) for the synthetic data. There are a few stars below (see discussion above) but none above the normality line. Therefore, one can be confident that fore- and background stars cannot mimic a statistically significant number of peculiar globular cluster members in the a versus $(g_1 - y)$ diagram. However, some negative outliers of NGC 104 and NGC 7099 (Fig. 57) can be caused by field stars.

NGC 104: no significant deviating Δa values for HB stars were found. About 10 outstanding objects are probable non-members and can be easily identified in the V versus $(g_1 - y)$ diagram.

NGC 6205: this is the most detailed investigated globular cluster among the targets. Three HB stars listed in Behr (2003a) were measured, namely, WF2-820 (No. 147), WF-2692 (No. 227), and WF2-3035 (No. 74). The latter shows no chemical peculiarities, whereas the other two have large overabundances of almost all iron peak elements from 1 to 1.5 dex compared to the cluster metallicity. The Δa values are perfectly in line with the abundances. For WF2-3035 an insignificant value of +8 mmag was found, whereas the other two stars were detected with +57 and +60 mmag. This lends to confidence that the Δa photometric system is indeed capable to detect chemically peculiar HB stars. However, further observations of such objects have to prove this conclusion. There are also several BHB stars which are below the normality line. For none of these objects, membership probabilities are available in the literature (Johnson & Pilachowski, 2012). One may speculate that this behaviour could be due to photometric variations. Such a behaviour is a common phenomenon for CP stars (Sect. 5), but was never investigated for members of globular clusters, yet.

As next step, the RGB stars published by Johnson et al. (2005) among the sample were investigated. In total, 10 stars were found in common. The [Mg/Fe] values for those objects range from −0.15 to +0.30 dex. None of them exhibit a significant Δa value, probably because the effect of the Mg

[4] http://www.univie.ac.at/nemo

lines compared to Fe for such rather low peculiarities is too small in the 5200Å region.

NGC 7099: there are no detailed elemental abundances for members in the literature available. There are three HB stars with a Δa detection, from which one object lies significantly above the HB in the V versus $(g_1 - y)$ diagram. If it is a member then this is probably a very interesting object for follow-up observations. The reasons why several fainter stars deviate from the normality line (as also seen for NGC 104), are not straightforward to determine without any additional observations. However, from the previous considerations one can think of non-members or very strong peculiarities of magnesium.

19 The Δa observations within the Large Magellanic Cloud

Our Galaxy's brightest satellite systems are the Large and Small Magellanic Clouds (LMC and SMC), which are only visible from the southern hemisphere. They are both classified as irregular dwarf galaxies and have been greatly distorted by tidal interaction with the Milky Way as they move close to it. It is even possible to observe streams of neutral hydrogen connecting them to the Milky Way and to each other. The significant astrophysical differences between the Magellanic Clouds and the Milky Way are (Choudhury et al., 2016)

- Different global structure including rotational/kinematical properties and magnetic field strength
- A factor of ten lower total mass
- Gas-rich
- Their overall metallicity is by a factor of two to four lower

Therefore, the formation and evolution of the Magellanic Clouds is significantly different to our Milky Way. It is possible to study the most important astrophysical mechanisms (convection, diffusion, accretion, rotation, and pulsation) in a completely different global environment. Since the construction of several major observatories in Chile was mainly argued with the exploration of the Magellanic Clouds, our knowledge about these galaxies was significantly enhanced in the last 40 years.

There are several important issues which can be answered from the observational side by establishing the incidence of CP stars in the LMC. Most important is the question if there is an influence of metallicity on the (non-)presence of peculiarities since metallicity seems to distinctly influence the star formation scenario in the sense that the formation of larger clusters is only possible in low metallicity media. It seems worthwhile to investigate whether the formation of magnetic peculiar and (non-magnetic) λ Bootis stars occurs in the same proportion to "normal" stars for all degrees of metallicity. Furthermore, one should ask whether different general magnetic field strengths in the surrounding area of star formation will lead to the same frequency of magnetic stars. While one could try to find a relationship concerning metallicity also in our own Galactic disc, the Magellanic Clouds offer a scenario with distinctly lower metallicity and lower magnetic field strength. In order to explain the characteristics of CP stars, stellar models (Sect. 14) are needed which take into account different metallicities, ages and magnetic field strengths.

Here, Δa observations of four fields including four star clusters (NGC 1711, NGC 1866, and NGC 2136/7) within the LMC are presented. In addition, the incidence of CP stars is estimated.

19.1 Target fields and observations

The observations were done on several nights at two different telescopes with the identical Δa filter set (Table 34):

- CASLEO: 215 cm telescope, TEK-1024 CCD, field of view of about 9.5', November 1998, August 2001 and January 2003, observer: O.I. Pintado
- ESO-LaSilla: Bochum 61 cm telescope, Thompson 7882 CCD, 384x576 pixels, 3'x4', April 1995, observers: H.M. Maitzen and E. Paunzen

In the following, a brief description of the observed bulge field and star clusters (Table 35) are given. Furthermore, comments are given for the surrounding field of NGC 1711.

Field 1: is located in the bulge of the LMC on its eastern edge (Cioni et al., 2000) centred at $\alpha = 04:45:20$ and $\delta = -69:12:00$ (2000.0). There are no prominent clusters or H II regions within this field in order to ensure that only the field population of the LMC is observed.

Field 2: it consists of a MS similar to that of NGC 1711 and an older population with a vertical extension of the red clump. Such a characteristics was observed in several areas of the LMC (Dolphin, 2000). It seems

Table 34: Characteristics of the used filters.

Filter	λ_C [Å]	Bandwidth [Å]	Transmission [%]
g_1	5027	222	66
g_2	5205	107	50
y	5509	120	54

Table 35: Observing log

	$\alpha(2000)$	$\delta(2000)$	Date	$\#_N$
Field 1	04 45 20	−69 12 00	11/1998	25
			08/2001	32
NGC 1866	05 13 40	−65 27 49	11/1998	76
NGC 1711	04 50 37	−69 59 01	01/2003	45
NGC 2136/7	05 53 07	−69 29 15	04/1995	12
			08/2001	29
			01/2003	28

that stellar evolution was rather uniform throughout the LMC until 200 Myr ago. Since then, a significant increase can be inferred from observational data. The different populations are also seen in Fig. 62 especially for the surrounding of NGC 1711.

NGC 1866: this is one of the most massive young clusters of the LMC in the age range 100 – 200 Myr (Marconi et al., 2013) and has been the subject of many studies. It is located in the outskirts of the LMC, so that field star contamination is not severe. This cluster contains many Cepheids which were used to determine the distance (Molinaro et al., 2012). The metallicity, $[Fe/H] = -0.55$ dex is typical for the LMC, the reddening $E(B-V) = +0.07$ mag is a little bit less than the mean value of the LMC.

NGC 1711: the only recent detailed investigation of this cluster was done by Dirsch et al. (2000) who derived the following parameters within the Strömgren photometric system: $\log t = 7.70(5)$, $[Fe/H] = -0.57(17)$ dex. However, these data are neither available in electronic form nor upon request from the authors. Therefore, it was not possible to testify their results. Kubiak (1990) lists an age of $\log t = 7.3$ on the basis of solar abundant isochrones and Johnson BVI photometry. Mateo (1988), on the other hand, derived an age between $\log t = 7.3$ and 7.7, respectively. This is in line with the results by Popescu et al. (2012) of $\log t = 7.3$ and 7.61 based on previously published broadband photometry and a stellar cluster analysis package.

NGC 2136/7: Dirsch et al. (2000) concluded that NGC 2136 and NGC 2137 are a physical binary (possibly triple) cluster system of same age ($\log t = 8.0$) and metallicity ([Fe/H] = −0.55 dex). The estimated reddening of $E(B-V) = 0.1$ mag is typical for other clusters in the LMC. As for NGC 1711, it was not possible to use the Strömgren photometry from Dirsch et al. (2000) for the calibration of the photometric values and the identification of objects because it has been available neither in electronic form nor upon request from the authors of the given reference. Furthermore, no printed tables are available. Mucciarelli et al. (2012) used a sample of high-resolution spectra to derive the kinematical and chemical properties of NGC 2136 and NGC 2137. They found that the two clusters share very similar systemic radial velocities, and have also indistinguishable abundance patterns. They suggest that the two clusters are gravitationally bound and that they formed from the fragmentation of the same molecular cloud that was chemically homogeneous.

19.2 Reduction and calibration

Here, a brief summary about the reduction and calibration process for NGC 1711 and its surrounding field is listed. It can be immediately compared with those for Galactic open (Sects. 17.2.1 and 17.3) and Globular clusters (Sect. 18.1).

For both fields, the results of the extensive $UBVR$ CCD survey conducted by Massey (2002) were used to calibrate and check the data. He presented accurate measurements for approximately 250 000 objects brighter than 18th magnitude in the SMC and LMC. Although there are several other sources of photometric data available, only data of Massey (2002) were included in the analysis in order to guarantee a sample free of bias on the basis of one widely accepted photometric standard system.

As mentioned before, it was tried to obtain the Strömgren $uvby\beta$ photometry from Dirsch et al. (2000) published for NGC 1711 and five other clusters in the LMC. But it is neither available in electronic form nor upon request from the authors.

For the calibration of the y and $(g_1 - y)$ to standard Johnson V as well as $(B-V)$ values, all stars in common with Massey (2002) were chosen.

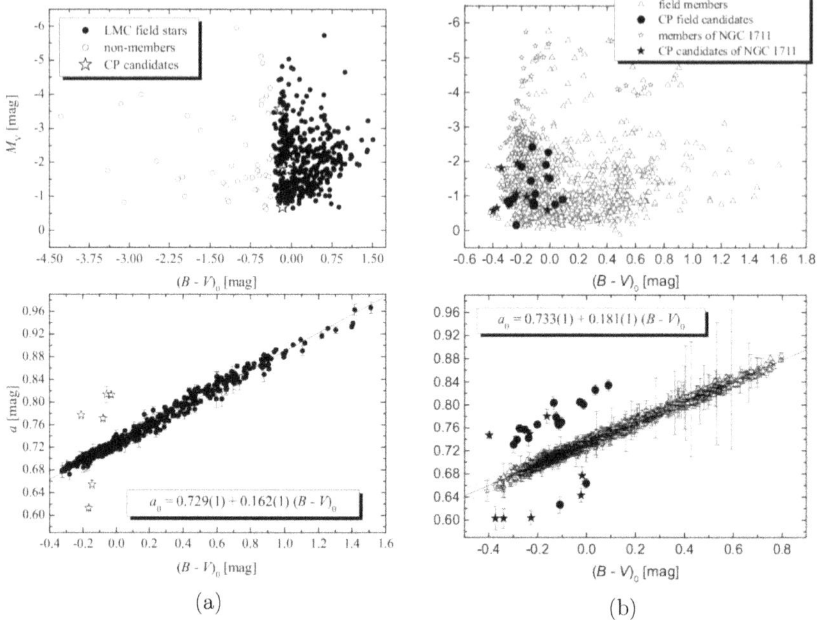

Figure 62: Observed diagrams for two program fields in the LMC. The solid line is the normality line whereas the dotted lines are the confidence intervals corresponding to 99.9%. The error bars for each individual object are the mean errors. The measurement errors of M_V are much smaller than the symbols and have been omitted. The scales for the upper and lower diagrams are different because the relevant range for peculiar objects in the a versus $(B-V)_0$ diagrams seemed worthwhile to be shown. The M_V and $(B-V)_0$values were calculated using a distance modulus of 18.5 and a reddening $E(B-V) = 0.1$ mag.

This procedure gives a sample of 372 objects. A linear least square fit resulted in:

$$V = -2.67(9) + 0.88(2) \cdot y$$

$$(B-V) = 1.24(6) + 1.44(8) \cdot (g_1 - y)$$

In parentheses are the errors in the final digits of the corresponding quantity.

A distance modulus of 18.5 mag for the LMC was published by Alves (2004) who reviewed and summarized all relevant measurements as well as their methods. Pietrzyński et al. (2013) derived a distance modulus $(m-M) = 18.49(5)$ mag and $A_V = 0.25$ mag as mean values for the LMC .

For the estimation of the reddening towards the observed fields, the maps of Oestreicher et al. (1995) were taken. Both fields are located in areas with $0.05 < E(B-V) < 0.12$ mag. This is in line with the results for NGC 1711 by Dirsch et al. (2000) who derived $E(B-V) = 0.09(5)$ mag on the basis of Strömgren *uvby* photometry. Therefore $E(B-V) = 0.1$ mag was adopted for both fields.

The adopted reddening value and distance modulus were checked by applying appropriate isochrones (Sect. 13) to the photometric measurements. The results for NGC 1711 are in good agreement with those from Kubiak (1990) and Dirsch et al. (2000).

For the normality lines, the $(g_1 - y)$ measurements were converted into $(B-V)$ values and dereddened. For the field of NGC 1711 and its surrounding, the following correlations were found, respectively:

$$a_0 = 0.716(1) + 0.162(1) \cdot (B-V)_0$$

$$a_0 = 0.733(1) + 0.181(1) \cdot (B-V)_0$$

The results are shown graphically in Figure 62. The 3σ limit for both fields is at ± 0.012 mag. The non-members in the surrounding field of NGC 1711 are objects which are much too blue in respect to the MS. One can only speculate about their nature since no independent $(B-V)$ values are available in the literature. The most probable explanation is emission in the g_1 filter which would yield such a blue colour index. Some of these object might also be very distant galaxies.

Because the observations were performed in the years 1998, 2001 and 2003, the behaviour of the detector and the intrinsic transformations of all data were checked and no significant trends or deviations from one year to the other were found.

19.3 The incidence of the apparent peculiar objects

A summary of the results from the CCD Δa photometric measurements in the LMC is given in Table 36. In total, 2742 objects of which 35 exhibit significant positive whereas 16 have negative Δa values (Table 36). The latter are either Be/shell or metal-weak objects and have, in general, no strong magnetic fields.

It has to be emphasized that in Sect. 14.1 it was shown on the basis of a synthetic Δa photometric system that only a shift of the normality line by about -3 mmag assuming an average metallicity of $[Fe/H] = -0.5$ dex relative to those in the solar neighbourhood occurs. The absolute Δa values for CP stars are not affected.

Objects with significant negative Δa values: this group consists of either classical emission line or metal-weak objects (like λ Bootis stars). In total, 16 of such objects were detected. For one object, the Be nature was spectroscopically confirmed (Sect. 19.4). Without any further photometric or spectroscopic data, no clear decision about the true nature can be drawn. However, it is not surprising to find a rather significant amount of these objects in NGC 1711 since it is a very young cluster still connected with stellar activity and emission of all kinds. Metal-weak MS stars in the relevant spectral range are very scarce (compared to metal-strong objects) in the field population of the solar neighbourhood (only a maximum of 2% of all stars, Sect. 6) and almost absent in open clusters older than $\log t = 7.30$ in our Milky Way (Gray & Corbally, 2002). This might be caused by accretion from a diffuse interstellar cloud together with diffusion as most probable mechanism for the λ Bootis phenomenon (Kamp & Paunzen, 2002). This scenario works at any stage of stellar evolution as soon as the star passes a diffuse interstellar cloud. Its dust grains are blown away by the stellar radiation pressure, while the depleted cloud gas is accreted onto the star. This would naturally generate an abundance pattern as found for the λ Bootis group, namely surface underabundances of most Fe-peak elements and solar abundances of lighter elements (C, N, O, and S). Since denser clouds within clusters dissipate on time scales of about 50 Myr, no λ Bootis stars are expected in older associations which make their appearance in clusters of the LMC rather unlikely.

The incidence of Be/Ae/shell and metal-weak stars in the LMC: these are the only groups which exhibit significant negative Δa values (Sect. 11) . The Be/shell population in the LMC is well studied (Keller et al., 1999). They found a fraction of MS stars that are Be objects between 0.1 and 0.34 (mean value of the Milky Way is about 0.17) varying for six very young clusters and their field population in the LMC. No correlation of the incidence with the age or metallicity was detected. Gray & Corbally (2002) concluded that the frequency of λ Bootis stars in the Galactic field among main sequence A to F objects is about 2% (no investigation for the LMC is available up to now). The age distribution of both groups are rather different. Whereas Be/Ae/shell stars are preferably found at young ages with $\log t < 7.5$, the members of the λ Bootis group in the Milky Way are evolved with a peak at $\log t = 9.0$ (Paunzen et al., 2002a). The detection capability of the Δa photometric system for a 3σ limit of 0.012 mag is at 50% as well as 10% for the λ Bootis and Be/shell group, respectively (Sects. 11.7 and 11.9). This means that 5% of the complete Be/Ae/shell as well as 60% of the λ Bootis population can still be detected with a high statistical significance at a limit of 0.012 mag. If one takes the maximum percentage of

Table 36: Statistics of observed stars in the spectral range up to F2 divided into NGC 1711, NGC 1866, NGC 2136/7, Field 1 (centred at $\alpha = 04{:}45{:}20$ and $\delta = -69{:}12{:}0$; 2000.0), Field 2 (surrounding field population of NGC 1711), Field 3 (surrounding field population of NGC 1866), and Field 4 (surrounding field population of NGC 2136/7). N_{all} is the total number of measured objects within the spectral range up to F2 or $(B-V)_0 = 0.3$ mag, N_+ and N_- are the numbers with significant positive as well as negative Δa values, N_1: $100 \cdot \frac{N_+}{N_{\rm all}}$,$N_2$: $100 \cdot \frac{N_-}{N_{\rm all}}$.

Name	$N_{\rm all}$	N_+	N_-	N_1	N_2
NGC 1711	109	3	5	2.8	4.6
NGC 1866	261	4	3	1.5	1.1
NGC 2136/7	73	3	3	4.1	4.1
Field 1	331	4	2	1.2	0.6
Field 2	622	15	2	2.4	0.3
Field 3	1239	4	–	0.3	–
Field 4	107	2	1	1.9	0.9

the relevant group members from Keller et al. (1999) and Gray & Corbally (2002), an upper frequency of 2.8% of objects with negative Δa values (1.7% Be/Ae/shell as well as 1.1% λ Bootis) among all main sequence B- to F-type stars should occur in the LMC for the derived detection limit of 0.012 mag. The mean and median value for all detected stars (Table 36) is 1.9 and 1.0%, respectively. These numbers are in excellent agreement with each other. The presence of these stars is, in general, depending on denser interstellar clouds (λ Bootis group) as well as stellar activity which is closely correlated with the age. However, it shows that Δa photometry, taking into account a certain detection limit, is able to find essentially the same frequency of Be/Ae/shell and metal-weak stars as other photometric as well as spectroscopic surveys.

The incidence of magnetic CP stars in the Milky Way: the detected 35 objects are most certainly magnetic CP (CP2 and/or CP4) stars (Sect. 6). Only for the four objects cooler than A0, $(B-V)_0 > 0$ mag, i.e. one can assume a membership to the CP2 group. Two aspects have to be taken into account when comparing the number of CP stars in the Milky Way and the LMC. First of all, the metallicities in the solar neighbourhood and the galactic open clusters are significantly higher than in the young

associations in the LMC which hardly reach -0.5 dex. The second issue is about the evolutionary effect on the incidence of magnetic CP stars. The Hipparcos data suggest that the CP group behave just like apparently normal stars in the same range of spectral types occupying the whole width of the main sequence with kinematic characteristics typical of thin disk stars younger than about 1 Gyr (Gómez et al., 1998). This finding was supported by Pöhnl et al. (2003) who investigated four young open clusters (ages between 10 and 140 Myr) with known CP2 members establishing the occurrence of such objects at very early stages of the stellar evolution. However, Hubrig et al. (2000) challenged these results and claim that the distribution of the magnetic CP stars of masses below 3 $M_{\odot}$ in the HRD differs from that of the normal stars in the same temperature range at a high level of significance, magnetic stars being concentrated toward the centre of the main sequence band. In particular, they argue that magnetic fields are detected only in stars which have already completed at least approximately 30% of their main sequence life time. This somewhat discrepant result might be understood by the detectability of resolved Zeeman patterns which requires a specifically slow rotation. This gives preference to finding such objects in advanced phases on the main sequence band where rotational velocities have been decreased by the growth of stellar radii.
Abt (1979) investigated the dependence of the percentage of CP members in open clusters on their age. His sample, divided into magnetic Ap (Si) and Ap (Sr,Cr) stars, includes data for 14 galactic open clusters as well as associations, a rather small number if one takes into account that at least 2500 open clusters are known in our Milky Way (Dias et al., 2002). The frequency of Ap (Si) objects with $-1.3 < M_V < 1.4$ mag increases from 4% to 8% during $\log t = 7.0$ and 8.0, respectively, whereas the youngest Ap (Sr,Cr) stars are found at $\log t = 7.5$ (3% at $\log t = 8.0$). For, at that time, known peculiar stars (Osawa, 1965) of the Bright Star Catalogue, Abt (1979) gives a number of 6.5% for the same magnitude and declination range as the investigated members of open clusters. No value for the CP4 group is listed in Abt (1979). Another source for a statistically analysis of galactic CP field stars is the conference paper by Schneider (1993) who considered also the Supplement to the Bright Star Catalogue and the CP stars listed in Renson (1991). He lists for the magnitude range $V < 7.1$ mag a lower limit of 16% (CP2 and CP4) and 5% (CP2 only) for stars hotter and cooler than A0, respectively. These numbers should be taken with care since Schneider (1993) quotes that "the single values should not be taken too serious because the MK classification influences the single values strongly". From the mentioned sources we conclude that one would expect an incidence for magnetic CP stars of at least 6% over the relevant spectral

range up to F2 main sequence objects. This number has to be treated as statistically lower limit for the percentage of magnetic CP stars for any test sample of main sequence objects in the Milky Way. The only limitation might be very young open clusters and associations ($\log t < 7.0$) where the incidence seems slightly less than 6% (Borra et al., 1982).

The incidence of magnetic CP stars in the LMC: first of all, one has to correct the observed ratio of CP stars for loss of objects because of the detection limit. In Sect. 11, a detection rate of 90% for CP2 and 65% for CP4 objects for a limit of 0.012 mag was estimated. However, there a factor of ten more CP2 than CP4 stars known in the solar neighbourhood known (Table 10). This ratio has to be assumed also valid for the LMC because no other sound measurement is available. Incorporating these facts for the values listed in Table 36, a mean and median value of about 2.7% for the incidence of magnetic CP stars in the whole spectral range up to F2 is inferred. This number is based on the measurements of 2742 individual objects. The occurrence of magnetic CP stars in the LMC is therefore only about half the value as in the Milky Way.

19.4 The spectroscopic verification of one CP and Be star

Let's recall that in Sect. 19, the results of the Δa system for fields in the LMC are presented. Furthermore, the incidence of CP stars in the LMC with that of the Milky Way is analysed (Sect. 19.3). However, all these investigations are based on one important assumption: classical CP stars do exist in the LMC and are unambiguously detected by Δa photometry.

It is therefore vital, to confirm these photometric results on the basis of spectroscopy. Here, spectroscopic observations made at Las Campanas Observatory are presented. They confirm one classical B8 Si star among the photometric sample, as well as one early B-type emission-line star which was also initially detected by its significantly deviating Δa value.

19.4.1 Target selection, observations, and reductions

From the stars with significant positive Δa values, detected as described in Sect. 19.1, the CP candidates #687, #1005, and #1093 were chosen. In addition, the metal-weak/Ae/Be/shell candidate #257 was selected for this study. The basic information for these stars are listed in Table 37. The V magnitudes of the targets, all located in the field of NGC 1866, are between 16.9 and 18.2 mag, near the limiting brightness for achieving usable spectra in moderate exposure times with the instrumentation available.

Table 37: Targets for the spectroscopic verification. X = Y = 0 corresponds to $\alpha(2000) = 05^h13^m40^s$ and $\delta(2000) = -65^o27'49"$. Notation for stars sorted after X and Y, respectively and according to Testa et al. (1999). σ_1 is the σ-value outside the corresponding confidence interval of the normality line. In parentheses are the errors in the final digits of the corresponding quantity.

No_1	No_2	X	Y	$(g_1 - y)$	a	Δa	σ_1	y	V	N
257		−153.15	−184.67	−0.716(3)	0.367(4)	−0.058	10.7	21.606(8)	17.059	76
687	4986	−44.75	−38.01	−0.804(8)	0.488(6)	+0.044	4.9	21.459(15)	16.908	60
1005	9527	+33.49	−47.05	−0.864(6)	0.477(3)	+0.045	9.6	22.308(4)	17.780	71
1093	9448	+52.65	+28.10	−0.932(8)	0.467(6)	+0.048	5.7	22.763(7)	18.248	71

The observations were performed with the Magellan Inamori Kyocera Echelle (MIKE), a high-throughput, double echelle spectrograph (Bernstein et al., 2003) on the 6.5-m Magellan telescope (Clay) at Las Campanas (Chile) on the nights of 18-20 November 2004 by Don J. Bord. The spectrograph employs a dichroic filter with separate blue and red fibres to provide simultaneous coverage of wavelengths spanning the range from 3200 to 10 000Å with a typical one-pixel resolution of 0.04 to 0.05Å. The spectra were binned 2x2 to produce a resolution of about 0.08 Å in the blue. This value comports well with the measured half-widths of sharp, unblended comparison lines in the ThAr spectrum near 4000Å which yields on the average 0.12Å. Integration times between 30 and 60 minutes were chosen to control the intensity of the night sky lines and to limit spectral contamination from neighbouring stars in crowded fields caused by rotation of the slit on the sky.

All reductions were performed with standard IRAF routines. After the bias, dark, and flat field corrections, the spectra were wavelength calibrated (including a heliocentric correction) using two ThAr standard spectra observed before and after the corresponding target. As a final step, the spectra were co-added to increase the signal-to-noise ratio.

19.4.2 Results and conclusions

Due to the faintness and the high spectral resolution of the instrument, the signal-to-noise ratios of the final spectra are very low, typically around 15. For the bluest and reddest regions of the spectra, no signal at all was observed because of the drop of the quantum efficiency of the CCDs.

For all stars, the hydrogen lines are the most prominent features in

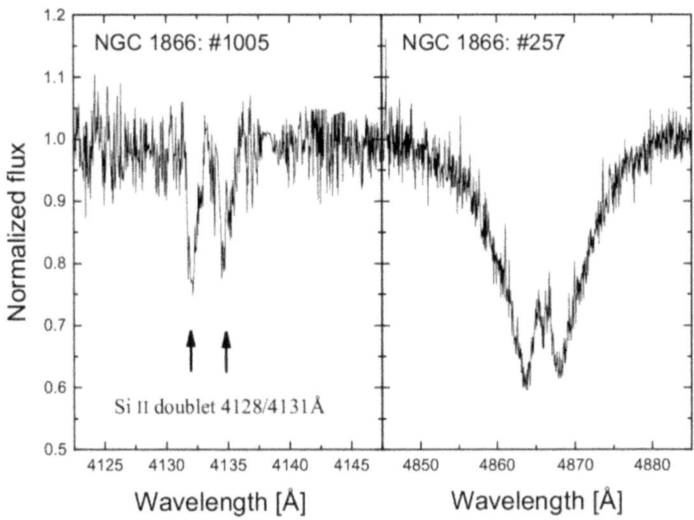

Figure 63: The Si II doublet detected in star #1005 (left panel) and the emission structure of Hβ in star #257 which exhibits a significant negative Δa value (right panel). The spectra have been shifted to the rest frame of the Sun. To within the measurement uncertainties (∼15%), the equivalent width of each Si-line is the same, viz. 0.30Å.

the spectra. The photometric candidates are thus clearly in the expected effective temperature range for classical CP stars and not contaminated by any late-type fore- or background objects.

One of the most prominent features of early-type CP stars is the Si II doublet at 4128/4131Å which is very much stronger than comparable lines in normal stars (Sect. 6). The enhancement of this doublet is an unambiguous characteristic of magnetic CP stars in general, and so it was primarily searched for this feature in the spectra.

For targets #687 and #1093 it was not possible to detect any features beside the hydrogen lines due to the low signal-to-noise ratio of the spectra and moderate rotational velocities of the stars. Applying several smoothing techniques (boxcar, FFT, and median) still revealed no additional features above 5σ of the mean noise level. Unfortunately, it was not possible to reliably analyse the 5200Å region because the depths of the flux depressions for the measured Δa values for these stars are too shallow for the low signal-to-noise ratio (Sect. 14.5.3).

The rotational speeds of these two stars can be estimated, however, from comparisons of the observed hydrogen line profiles with synthetic spectra. Establishing a credible continuum level for the hydrogen lines in the echelle spectra is neither easy nor unambiguous, and normalization was carried out using the shape of the flat field function of the corresponding aperture. From different numerical experiments using a range of stellar parameters, it was found that the most probable $v \sin i$ values for #687 and #1093 fall within the range of 100 to 150 km s^{-1}.

Figure 63 shows the Si II doublet observed for #1005 with the most significant Δa value detected, 10.8σ above the normality line. The separation of the two lines was measured which is, within the errors, identical with the theoretical one. The spectral feature is so strong that one immediately classifies this object as a typical silicon star of the CP2 group. The equivalent widths of the doublet were measured to be 0.30 Å for each line with an error of about 15%. For normal late B-type stars (dwarfs and giants), the corresponding value is 0.19 Å (Jaschek & Jaschek, 1995). Notice that the equivalent width of the Si II doublet for dwarfs is at a maximum for spectral type B8 and drops to 0.15 and 0.11 Å for B7 and B9, respectively. The corresponding equivalent widths for late B-type CP stars range from 0.2 to 0.4 Å, depending on the rotational phase and the strength of the stellar magnetic field (López-García & Adelman, 1999). These values are in agreement with the measurements of #1005.

Testa et al. (1999) published $V = 17.797(5)$ and $(B-V) = 0.014(13)$ mag for this star. Using the distance modulus of 18.33 and $A_V = 0.25$ mag for NGC 1866 and its surrounding taken from Salaris et al. (2003), one derives an absolute magnitude of -0.8 mag which coincides with a main sequence late B-type star. Within the given errors one can classify #1005 as B8 Si star. From the original as well as a smoothed spectrum, the upper limit of $v \sin i$ as 50 km s^{-1} from the hydrogen lines and Si II doublet was determined.

The emission characteristics of Hβ for star #257, which exhibits a significant negative Δa value at a 10.7σ significance level, is shown in Figure 63. This object is about 0.7 mag brighter that #1005 which suggests an early B-type. Again, this result is in agreement with the findings for similar stars in the Milky Way.

From the wavelength calibrated spectra (Fig. 63), using the hydrogen lines, it was also possible to obtain estimates of the heliocentric radial velocities for each target star. The final values are between 280 and 320 km s^{-1} with errors of about 20 km s^{-1}. This clearly establishes them as true members of the LMC and not Galactic foreground objects (Prevot et al., 1985).

These two examples show that the Δa photometric system is indeed able to distinguish CP and Be from normal type objects via the classical tools (Sect. 3). However, a certain contamination from other objects, for example Galactic foreground stars, can not be excluded. To investigate this possible effect, more spectroscopic observations of the bona-fide photometric candidates are needed. For this, at least 8 meter class telescopes are needed, even employing classification resolution spectroscopy.

19.5 The (non-)variability of magnetic CP candidates in the LMC

The variability of magnetic CP (mCP) stars is explained in terms of the oblique rotator model (Sect. 6), according to which the period of the observed light, spectrum, and magnetic field variations is the rotational period. The photometric changes are due to variations of global flux redistribution caused by the phase-dependent line blanketing and continuum opacity namely in ultraviolet part of stellar spectra (Krtička et al., 2012).

The amplitude of the photometric variability is a combination of the characteristics of the degree of nonuniformity of the surface brightness (spots), the used pass band, and the line of sight. The observed amplitudes are up to a few tenths of magnitudes. However, for some stars one also fails to find any rotational induced variability at all. The locations of spots used to have connection with the dipole-like magnetic field geometry (Krtička et al., 2009).

Besides the confirmation by spectroscopic data (Sect. 19.4.2), it is important to analyse if the bona-fide CP candidates (Table 36) also show the same variations as their Galactic counterparts which would be a direct indicator for the presence of a stellar magnetic field.

The list of mCP candidate stars were compared with the OGLE database for corresponding measurements on the basis of equatorial coordinates and V magnitudes. After a first query, the positions in the original images and those of OGLE were inspected by eye. In total, fourteen matches in both sources were found. The final list of stars is given in Table 38.

A further analysis revealed that the sample of mCP candidates is contaminated by two late-type stars. The first, 135.3 4273, was recognised as the short-period Cepheid OGLE-LMC-CEP-0327 pulsating in its first overtone with the period $P = 0.8821659(13)$ d (Soszyński et al., 2008). The second, 135.3 30107, is a normal, non-variable K-type giant. These objects were adopted as control stars and their photometric data were analysed together with other stars of the sample. The HRD of all studied stars constructed from data given in Table 38 is shown in Figure 64.

Table 38: The basic photometric parameters and the photometric characteristics of CP candidates in the V and I bands. $\overline{V}$ is the mean magnitude in V filter, $\overline{M_0}$ the V absolute magnitude assuming the distance modulus $m - M = 18.49(5)$ and $A_V = 0.25$ mag (Sect. 19.2), $(\overline{V-I})$ the mean colour index, the mean Δa index, $N_{\rm V}$ and $N_{\rm I}$ the numbers of V and I observations, δm_V and δm_I the typical uncertainties of particular measurements, $1/P$ the frequency of the maximum effective amplitude of $A_{\rm m}$ or S/N, while $A_{\rm m}$ and $A_{\rm ms}$ are the observed maximum modified amplitude and expected maximum modified amplitude. S/N and S/N_s are the maximum signal-to-noise value and its expected value; "prob" means the probability of the reality of designate frequency of periodic variations.

No.	OGLE LMC +	$\overline{V}$ [mag]	$\overline{M_0}$ [mag]	$(\overline{V-I})$ [mag]	Δa [mag]	N_V	δm_V [mag]	N_I	δm_I [mag]
1	135.3 4273*	17.80	−0.95	+0.85	(0.089)	67	0.013	370	0.013
2	135.3 30107*	17.76	−0.98	+1.36	(0.094)	65	0.011	466	0.010
3	136.7 861	17.60	−1.14	−0.04	0.078	28	0.010	301	0.018
4	136.7 16501	19.16	+0.41	+0.03	0.087	24	0.024	266	0.058
5	136.8 678	17.84	−0.90	+0.01	0.095	45	0.011	435	0.021
6	136.8 1801	18.79	+0.05	+0.11	0.085	45	0.019	437	0.041
7	136.8 1873	18.64	−0.10	+0.02	0.056	45	0.017	437	0.039
8	136.8 2002	18.67	−0.08	+0.10	0.054	45	0.018	434	0.038
9	136.8 3694	19.10	+0.35	+0.05	0.052	44	0.023	421	0.056
10	136.8 3875	18.98	+0.24	+0.04	0.087	43	0.022	434	0.052
11	190.1 1445	18.08	−0.65	−0.03	0.095	52	0.013	454	0.037
12	190.1 1581	17.89	−0.85	+0.01	0.044	51	0.012	454	0.031
13	190.1 2822	18.68	−0.06	+0.02	0.060	53	0.018	454	0.056
14	190.1 15527	17.16	−1.57	+0.01	0.041	53	0.008	453	0.019
		$1/P$ [d^{-1}]	$A_{\rm m}$ [mag]	$A_{\rm ms}$ [mag]	S/N	S/N_s	prob %		
1	135.3 4273*	1.1336	0.175	0.041	51	4.8	100		
2	135.3 30107*	0.6522	0.005	0.006	4.5	4.5	16		
3	136.7 861	1.0088	0.014	0.013	5.1	4.5	29		
4	136.7 16501	0.2286	0.044	0.040	5.1	4.6	28		
5	136.8 678	1.4388	0.013	0.013	4.7	4.6	18		
6	136.8 1801	1.1811	0.027	0.023	5.3	4.3	30		
7	136.8 1873	0.6549	0.021	0.022	4.2	4.3	20		
8	136.8 2002	0.4372	0.020	0.022	4.1	4.5	13		
9	136.8 3694	0.9838	0.031	0.033	4.4	4.7	23		
10	136.8 3875	1.8641	0.031	0.032	4.6	4.7	10		
11	190.1 1445	0.9183	0.017	0.016	5.6	5.3	26		
12	190.1 1581	0.8075	0.019	0.015	7.1	5.3	49		
13	190.1 2822	1.6884	0.022	0.024	4.6	5.1	11		
14	190.1 15527	2.0260	0.011	0.009	6.7	5.5	32		

*The control stars: the first is a short-periodic Cepheid, the second a non-variable K giant.

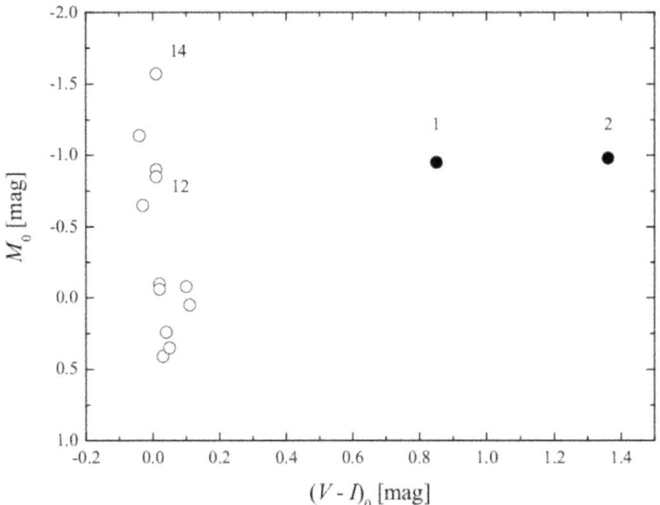

Figure 64: The HRD of all analysed stars. M_0 is the mean V absolute magnitude corrected for extinction $A_V = 0.25$ mag assuming the distance modulus $m - M = 18.49$; $(V - I)_0$ is the colour index corrected for the corresponding interstellar reddening by 0.1 mag. Open circles clustering towards the MS indicate mCP candidates, the filled circle denoted by 1 is the mean position of star 1 (135.3 4273), which is a short-period Cepheid. Star 2 (135.30107) is a normal non-variable K giant. Stars 12 and 14 are mCP candidates suspected for their rotationally modulated light variations.

The data of the mCP candidates contain the $\mathrm{JD_{hel}}$ date of measurement, t_i, the measured V and I magnitudes m_i, and the estimate of its internal uncertainty δm_i. It was found that the last quantity represents the real uncertainty of the magnitude determination quite well and that is why it was used to weight the individual measurements according to the relation $w_i \sim \delta m_i^{-2}$.

The mean V magnitudes of mCP candidates (stars 3 to 14) span the interval from 17.16 to 19.16 mag, respectively This corresponds to an interval of absolute V magnitudes from -1.4 to $+0.6$ mag. It corresponds mainly to Si-type of mCP stars (Sect. 6) and matches the results given in Section 19.4.2.

The median of the amplitude of light variations of mCP stars in V is about 0.032 mag (Bernhard et al., 2015a). The typical error of the determination of one I measurement of the mCP candidate set is $\sigma_{\mathrm{It}} \sim 0.038$ mag and the typical number of measurements is $N_{\rm It} \sim 440$ (Table 38). Consequently, the expected uncertainty of $\delta_{\rm It}$ of the amplitude determination of periodic variations can be estimated as $\delta_{\rm It} = \sqrt{8/N_{\rm It}}\,\sigma_{\rm It} \sim 0.005$ mag. The same quantities for V measurements are $\sigma_{\rm Vt} \sim 0.015$ mag and $N_{\rm Vt} \sim 45$, $\delta_{\rm Vt} = \sqrt{8/N_{\rm Vt}}\,\sigma_{\rm Vt} \sim 0.006$ mag.

Before starting the analysis of V and I photometric data, the expected relationship between infrared and visual variations of mCP stars were investigated in more detail.

19.5.1 *I* light variability of mCP stars

The search for periodic rotationally modulated variations in the sample of 12 LMC mCP candidates is based mainly on the analysis of the variability of these stars in the near infrared, while the Galactic mCP star variability is studied mainly in the visual region (Sect. 6). Unfortunately, the information about the infrared variability of Galactic mCP stars is very scarce. From the few observations available (Musielok et al., 1980; Catalano et al., 1998b; Wraight et al., 2012) one can conclude, that the amplitude of the light variations in the near infrared filters is similar but not identical to that in visual filters. This can be easily understood from the theoretical model of the light variability of mCP stars (Molnar, 1975; Krtička et al., 2007; Shulyak et al., 2010), which is based on the light redistribution from the shorter wavelength region (typically the far ultraviolet) to the longer wavelength region (typically the visual and infrared) because of enhanced opacity in the regions with overabundant elements. The infrared continuum lies in the Rayleigh-Jeans part of the flux distribution function, where the ratio of the two different fluxes is independent of wavelength.

To derive a quantitative prediction for the mCP star light variability in the near infrared, the successful models of the visual light variability of HD 37776 were employed ($T_{\rm eff} = 22\,000$ K, Krtička et al., 2007), HR 7224 ($T_{\rm eff} = 14\,500$ K, Krtička et al., 2009), and CU Vir ($T_{\rm eff} = 13\,000$ K, Krtička et al., 2012) and predicted the light curves in the I filter. The light curves are calculated from the surface abundance maps of these stars (Kuschnig et al., 1999; Khokhlova et al., 2000; Lehmann et al., 2007) using TLUSTY model atmospheres and the SYNSPEC spectrum synthesis code (Lanz & Hubeny, 2003, 2007). For these stars, no observations in I are available.

The light curves are predicted using specific intensities filtered by appropriate transmission curve. The OGLE transmission curve were fitted by a suitable formula with a precision better than 3% in the form of

$$M(\lambda) = \begin{cases} \exp\left(a_1 x + a_2 x^2\right), & \lambda < \lambda_0, \\ \exp\left(a_4 x + a_5 x^2 + a_6 x^3\right), & \lambda > \lambda_0, \end{cases} \tag{29}$$

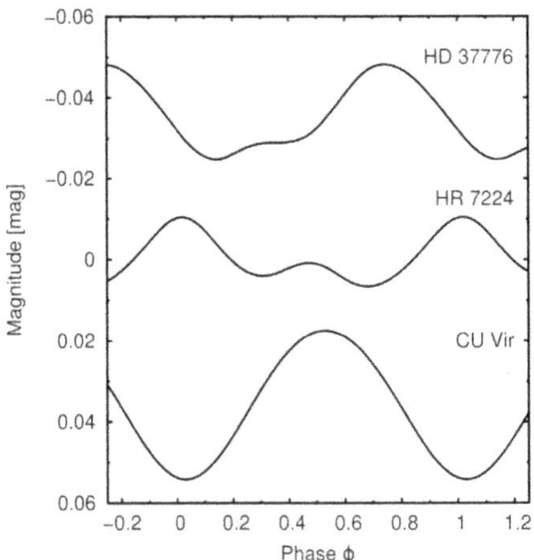

Figure 65: The predicted light curves in the I band of selected Galactic mCP stars.

where the variable x is connected with the wavelength λ in Å as

$$x = \left(\frac{\lambda - \lambda_0}{\sigma}\right)^2, \tag{30}$$

and

$$\lambda_0 = 8111\,\text{Å}, \qquad \sigma = 733\,\text{Å}, \tag{31a}$$
$$a_1 = -0.501, \qquad a_2 = -0.329, \tag{31b}$$
$$a_4 = -0.400, \qquad a_5 = +0.452, \qquad a_6 = -1.03. \tag{31c}$$

The resulting light curves are shown in Figure 65. From this plot one can conclude that if the LMC mCP stars have the same amplitude as their Galactic counterparts, their light variations should be detectable by OGLE (Table 38).

The detection of photometric variations of LMC mCP candidates is not straightforward because of their faintness. That is why their light behaviour was approximated by the simplest possible model assuming that light curves in all filters are simple sinusoidal with the same amplitude. Whenever a signal of this kind is detected, it is possible to improve the model even more.

19.5.2 Periodograms

The periods of variable stars are usually inferred by means of various kinds of periodograms. Two versions of periodograms were chosen based on the fitting of detrended measurements y_i with uncertainties σ_i done in the moments t_i by an ordinary sine-cosine model, $f(\omega,t) = b_1\,\cos(\omega\,t)+b_2\,\sin(\omega\,t)$, where $\omega = 2\pi/P$ is an angular frequency and P the period, using the standard χ^2 least-squares method. If the sum $\chi^2(\omega)$ is for the given ω minimum the so-called *modified amplitude* $A_{\mathrm{m}}(\omega)$ has to be maximum,

$$\sum_{i=1}^{N}\frac{y_i^2}{\sigma_i^2} - \chi^2(\omega) = \sum_{i=1}^{N}\left[\frac{f(\omega,t_i)}{\sigma_i}\right]^2 = A_{\mathrm{m}}^2\sum_{i=1}^{N}\frac{1}{8\,\sigma_i^2}, \quad\Rightarrow \tag{32}$$

$$A_{\mathrm{m}}(\omega) = \sqrt{\frac{8}{\sum\sigma_j^{-2}}\sum_{i=1}^{N}\left[\frac{b_1(\omega)\cos(\omega t_i)+b_2(\omega)\sin(\omega t_i)}{\sigma_i}\right]^2}, \tag{33}$$

where $b_1(\omega)$ and $b_2(\omega)$ are coefficients of the fit. When there is a uniform phase coverage, the modified amplitude is equal to the amplitude of sinusoidal signal. The best phase sorting of the observed light variations corresponds to the angular frequency ω_{m} with the maximum of modified amplitude $A_{\mathrm{m}}(\omega)$.

The second LSM type of periodogram uses for the significance of individual peaks a robust signal-to-noise (S/N) criterion, which is defined as

$$S/N(\omega) = \frac{Q(\omega)}{\delta Q(\omega)} = \frac{b_1^2(\omega)+b_2^2(\omega)}{\delta\left[b_1^2(\omega)+b_2^2(\omega)\right]}, \tag{34}$$

where $\delta Q(\omega)$ is an estimate of the uncertainty of the quantity $Q(\omega)$ for a particular angular frequency.

The properties of the $S/N(\omega)$ criterion were tested using thousand samples with sine signals scattered by randomly distributed noise. It was found that if there is no periodic signal in this data, the median of the maximum $S/N(\omega)$ value in a periodogram is 4.52; in 95% of cases, a S/N value between 4.2 and 5.4 was found. The occurrence of peaks definitely higher than 6 indicates possible periodic variations.

During the treatment of OGLE-III time series, it was concluded that both types of periodograms correlate very well with other time-proven estimates, the Lomb-Scargle (Horne & Baliunas, 1986) periodogram, for example. So one can consider them as generally interchangeable.

All frequencies from 0 to $2.1\,\mathrm{d}^{-1}$ were tested. The upper limit is slightly above the frequency of the fastest rotating CP star known to date (HD 98000, with the period of $P = 0.466\ \mathrm{d}^{-1}$, see Hümmerich et al., 2017).

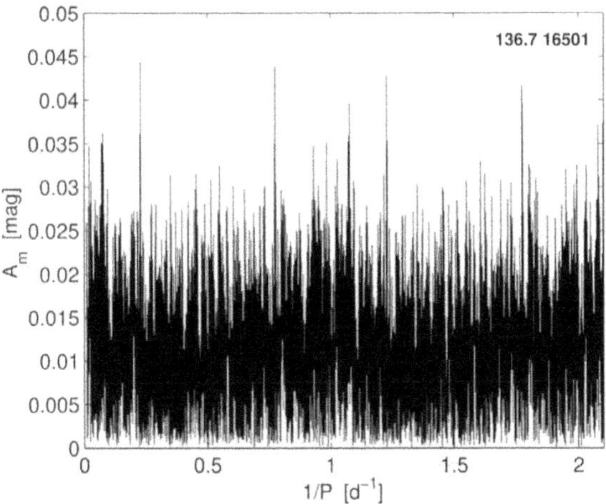

Figure 66: A typical periodogram of a mCP candidate, star 4 (136.7.16501); all peaks are inconspicuous.

19.5.3 Discussion of periodograms

The periodograms for all stars of the sample were constructed and frequencies $f = 1/P$ of their maximum peaks, modified amplitudes $A_{\rm m}$, and S/Ns for these peaks calculated. These quantities are given in Table 38. The periodograms of the LMC mCP candidates were compared with those of the control stars, where the variability of OGLE-LMC-CEP-0327 and its aliases conjugated with the period of a sidereal day (Fig. 67) were clearly detected. The results exactly agree with the previous determination of the period and other characteristics of the star. This lends confidence to the applied time series analysis method.

The periodograms of the known Cepheid and the mCP candidates plus the second control star apparently differ: the peaks here have only a very narrow margin to the body of a pure scatter (compare Fig. 66 with Fig. 67). There are additional arguments to explain why all periods except those found for the stars 12 and probably star 14, listed in Table 38, are very likely only coincidences:

- The distribution of the maximum amplitude frequencies is shallow and the found frequencies cover the studied interval 0 to 2.1 d^{-1} more or less evenly. This is a sign that observed peaks are most likely mere coincidences.

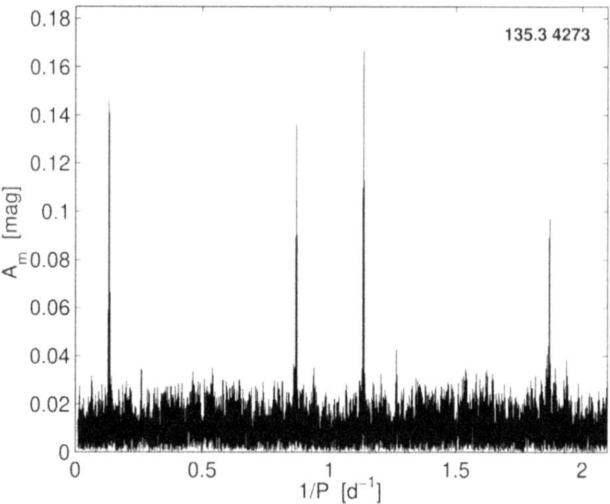

Figure 67: The basic period of the overtone-Cepheid OGLE-LMC-CEP-0327 (135.3 4273) plus its aliases conjugated with the period of a sidereal day.

- The amplitudes of light changes of Galactic mCP stars show no correlation with the colour index, spectral type, or absolute magnitude. It is in sharp contrast with observations of LMC mCP candidates whose observed maximum amplitudes exhibit strict monotonic dependence on the mean magnitude (see circles in Fig. 68). It suggests that the nature of observed variability of LMC objects might be different.
- It is conspicuous that the median of S/N ratios of the highest peaks in periodograms of individual CP candidates is only 4.7, corresponding to the median of maxima peaks $S/N = 4.52$ in the case of pure scatter. It indicates the lack of detectable periodic signal in the majority of inspected mCP candidates, with two exceptions (stars 12 and 14) which are discussed in the following.

Therefore, the hypothesis was tested that the distribution of the data is random via a heuristic *shuffling method*. The data of each mCP candidate was re-analysed in the same way as described above, only all the individual observations (magnitudes and their uncertainties) were randomly shuffled, the times of the observations remained the same. This should extinguish any periodic signal. As next step, A_{ms} and S/N_{s} for each shuffled version of data were calculated. The median of the both shuffled quantities is given

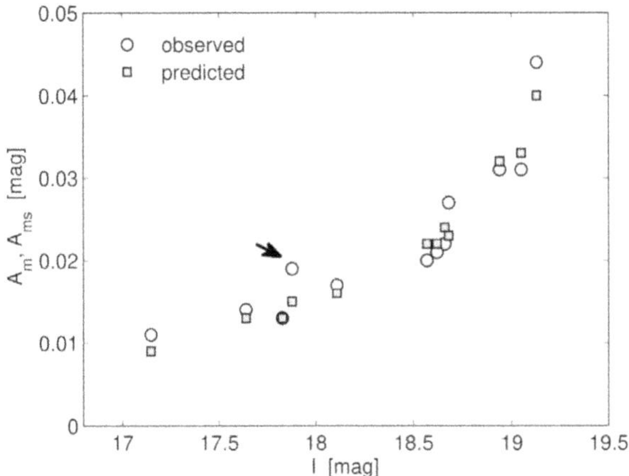

Figure 68: The dependence of the maximum modified amplitudes of the observed (A_{m} – circles) and the shuffled data (A_{ms} – squares) on the mean I magnitude of mCP candidates. The largest positive relative deviation is found for the star 12 (190.1 1581) and is denoted by an arrow.

in Table 38 and Figure 68. It is apparent that the results are nearly the same as in the case of the original, unshuffled data with two exceptions (stars 12 and 14).

As an added and independent test, we applied two different time series analysis methods, namely the modified Lafler-Kinman method (Renson, 1978) and the Phase-Dispersion-Method (Stellingwerf, 1978). All computations were done within the programme package Peranso (Paunzen & Vanmunster, 2016). Using these utilities only upper limits for variability were found, but no statistically significant detection except for the case of star 1. The frequencies discussed in the following for stars 12 and 14, are detected on a 3.4 and 3.9 σ level, respectively.

19.5.4 The mCP candidates suspected of periodic light variations

Both objects suspected of periodic light variations, star 12 (190.1 1581) and star 14 (190.1 15527), display relatively low amplitude light variation which are only slightly above the detectability by OGLE-III V and I photometry. Therefore, a very simple model of their behaviour assuming the linear ephemeris and the sine form of the V and I variations with different

Table 39: The derived parameters of the light variation model given in Equation 35.

		LMC 190.1 1581	LMC 190.1 15527
$M_0 - 2\,450\,000$		3710.895(36)	3710.245(12)
P	[d]	1.23836(5)	0.493570(8)
$\overline{m}_V$	[mag]	17.8917(16)	17.1698(10)
a_V	[mag]	0.022(4)	0.017(3)
$\overline{m}_I$	[mag]	17.8814(13)	17.1508(8)
a_I	[mag]	0.016(4)	0.008(2)

amplitudes was used. The following model of light changes were obtained

$$\vartheta(t) = \frac{t - M_0}{P}, \quad m_c(\vartheta) = \overline{m}_c - \frac{a_c}{2}\cos(2\,\pi\,\vartheta), \tag{35}$$

where $\vartheta(t)$ is a phase function, M_0 is the moment of the basic maximum, P is the period in days, $m_c(t)$ is a predicted magnitude in filter c ($c = V,\, I$) at the time t, $\overline{m}_c$ is a mean value of the magnitude in the filter c, and a_c is an amplitude in the filter c. The values for both stars derived by standard least-squares minimization regression are given in Table 39.

Star 12 (190.1 1581) is one of the brighter stars in the sample. Its mean absolute magnitude $\overline{M}_0 = -0.85$ mag and negative dereddened index $(\overline{V - I})_0 = -0.09$ mag indicates that this object belongs among the late B-type stars. If it is a true mCP star, it is most likely a Si- or He-weak type object (Sect. 6).

In its periodogram one finds a dominant peak at the frequency $f = 0.8075\ \mathrm{d}^{-1}$ corresponding to the period $P = 1.2433\ \mathrm{d}^{-1}$. The second peak of nearly the same height consists of two peaks at the frequencies $f = 1.8075\ \mathrm{d}^{-1}$ and $f = 1.81025\ \mathrm{d}^{-1}$. Their distance of $0.00275\ \mathrm{d}^{-1}$ corresponds to the reciprocal value of the sidereal year in days. The internal structure indicates that it is an alias of the basic period. The bootstrap test of the reality of the found period gives 49%, which is the absolute maximum among the mCP candidates in the sample. The V and I phased light curves of star 12 (190.1 1581) are shown in Figure 69.

A careful inspection of CCD images of this star reveals that it is a visual binary with a companion that is about 1.5 magnitude fainter than the target. The OGLE-III photometry was apparently done for both components. Future photometric observations and treatments of them should take this into account. On the other hand, the modest ratio of amplitudes of variations in V and I filters completely fulfils the theoretical expectation.

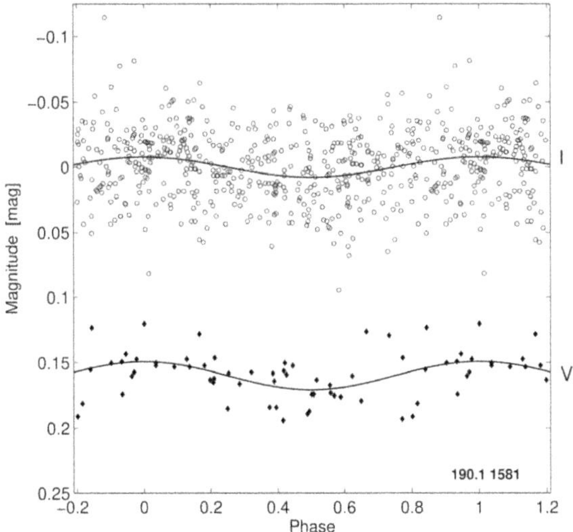

Figure 69: The V and I light curves of star 12 (190.1 1581) plotted according to the ephemeris given in Table 39.

Star 14 (190.1 15527) is the brightest and one of the hottest mCP candidates in the sample. With $M_0 = -1.57$ mag and $(V-I)_0 = -0.09$ mag it is a late Bp star with Si or He-weak chemical peculiarity (Sect. 6). The most striking result of its analysis is the found period of $P = 0.494$ d^{-1}, which is rather short. If it is confirmed, it would be one of the fastest rotating mCP stars detected so far.

Unfortunately, the found period shorter than 12 hours is not as firmly established as in the case of the previous mCP candidate with larger light changes. It results in the relatively low appreciation of the reality probability of the found period only 30%. The rather small variations in the I filter (Fig. 70) are also not in favour of variability, nor are the simulations with randomly distributed magnitudes that also constitute periodogramme peaks slightly larger than 2 and 1. That is why one suspects the feature to be an artefact caused by the specifically distributed observations.

In conclusion, it is suggest that the periodic variations of star 14 (190.1 15527) remains an open question.

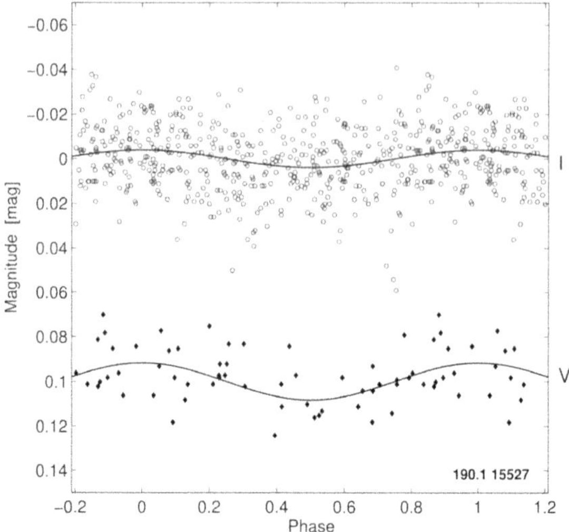

Figure 70: The V and I light curves of star 14 (190.1 15527) plotted according to the ephemeris given in Table 39.

19.5.5 Conclusions

The OGLE-III photometry for fourteen stars of which twelve are mCP candidates from Sect. 19.3 were analysed. In total, 6476 individual photometric measurements in V and I were used to perform a time series analysis. For this purpose, periodograms were calculated including a bootstrap-based probability of the statistical significance. Although the median of the amplitudes of maximum peaks, median $A_{\mathrm{m}} = 0.02$ mag in the periodograms of the mCP candidates, does not contradict the expectation for Galactic mCP stars, other circumstances suggest that the prevailing majority of these peaks might not be statistically significant. Periodic variations in only two of the mCP candidates were found, of which one is doubtful.

It seems that the rotationally modulated variability of the studied mCP candidates in the V and I bands is very weak (if present at all). It is very likely that the true amplitudes are much lower than 0.01 mag, which is the limit derived from the time series analysis of the brightest mCP candidate.

From this finding one is able to conclude that the spots on the stellar surfaces of these objects are not as contrasting as those of their Galactic counterparts; in other words, the overabundant elements (Si, Fe, Cr, and so on) are more evenly distributed over the surface or the maximum abundance is lower. This phenomenon has also been observed for non-magnetic Galactic Am or weakly magnetic HgMn stars.

The low contrast of LMC mCP photometric spots could be explained by their lower global magnetic fields strengths compared to those of their Galactic counterparts. It may relate with the lower level of interstellar magnetic field in the LMC if compared with the Galactic field, which results in the overall lower percentage of mCP star appearance compared to the Milky Way (Sect. 19.3).

As future steps, retrieving time series of the remaining published mCP candidate stars is suggested and getting more accurate measurements of the presented stars. In addition, spectroscopic observations would shed more light on the surface characteristics.

20 A case study of using Gunn g data to improve Δa

A case study was conducted for improving the existing Δa photometric system. It has been shown (Sect. 9) that the currently used filters are already optimized to measure the typical flux depression of magnetic CP stars.

The fundamental idea is to employ a broader band filter which measures the 5200Å region and the continuum blue- and redwards of it. The paradigm is the β index which consists of a narrow and wide filter centred at Hβ (Crawford, 1958). However, the β index alone is not able to give any unique astrophysical information. It always has to be used with an additional index like c_1 or m_1 (Strömgren, 1966). This strategy is also valid for this analysis.

The corresponding filter to upgrade the Δa photometric system has to fulfil the following characteristics

- The bandwidth should be not wider than 1000Å to avoid blurring the effect of the 5200Å region
- The central wavelength should be close to 5200Å
- It should measure the continuum blue- and redwards of the 5200Å region
- The contrast of CP and normal stars should be significant or at least larger than +15 mmag

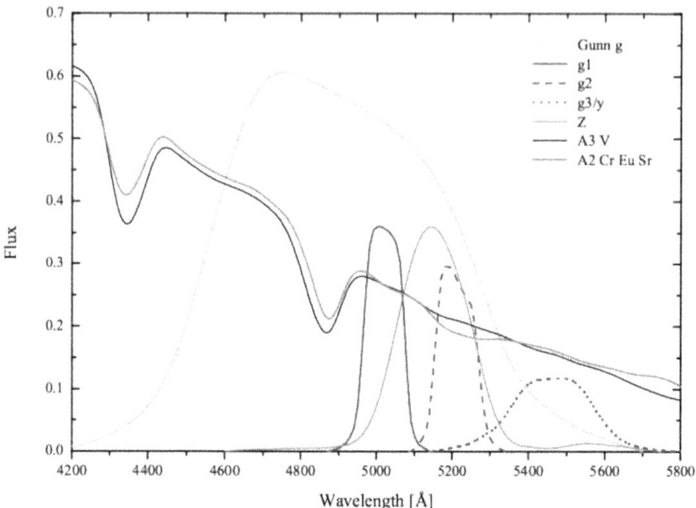

Figure 71: A comparison of the Gunn - g (green line), Δa (blue), and Vilnius Z (magenta) filters together with spectrophotometric data of HD 102647 (A3 V; black) and HD 118022 (A2 Cr Eu Sr; red) taken from Section 9.

- The effect of the luminosity should be minimized
- It should be widely available, well established, and tested

One of the choices would be the Z filter ($\lambda_c = 5160$ Å, FWHM $= 210$ Å) of the Vilnius $UPXYZVS$ photometric system (Straižys & Sviderskiene, 1972). More details of this system and its peculiarity indices are given in Section 4. There are two disadvantages: it does not measure the continuum redwards of 5200Å and it is hardly available at international observatories.

Finally, the g filter (Fig. 71) of the photometric system by Thuan & Gunn (1976) was chosen for a more thoroughly investigation. Its adopted version was used also for the Sloan Digital Sky Surveys (SDSS). This means that the filter is widely available at international observatories.

The synthetic Δa and g photometric grids were taken from the Vienna New Model Grid of Stellar Atmospheres, NEMO (Heiter et al., 2002b). The following parameter ranges were used: $6000 < T_{\mathrm{eff}} < 20\,000$ K (step size of 200 K for $T_{\mathrm{eff}} < 10\,000$ K, and 500 K for models hotter than that),

$2.0 < \log g < 5$ dex (step size of 0.2 dex) for [Fe/H] = [−2,0,+1], and microturbulent velocities $\xi_t = [0,1,2,4,8]$. This should cover giants and MS stars cooler than a spectral type of about B2. As next step, several combinations of the filters were investigated. The (g/g_2) versus $(g1 - y)$ diagram gives a null result, or in other words, the contrast of normal and peculiar objects is only ±3 mmag. From all different combinations, two parameters are worthy for a further consideration

$$a_1 = g_2 - g \tag{36}$$

$$a_2 = (g_1 - y)/(g - g_2) \tag{37}$$

Figures 72 and 73 show the results for three different metallicities [Fe/H] = [−2,0,+1], and a microturbulent velocities ξ_t of $2\,\mathrm{km\,s^{-1}}$. The left panels show the MS whereas the right ones also include the effect of luminosity. The blue asterisks denote the corresponding T_{eff} values in steps of 1000 K starting with 6000 K at the red end of $(g_1 - y)$.

One has to keep in mind that only distinct models have been calculated. Looking, for example at Fig. 72 in the upper left panels, for cool type models, "vertical points" are visible. This means that the models cover the whole area and overlap for the coolest temperatures.

First of all, no significant influence of different ξ_t values on these diagrams was detected. Notice that in Sect. 14.7, it was concluded that an increase of ξ_t from 2 to 4 km s^{-1} can raise synthetic Δa values by up to +10 mmag. This is caused by an decrease of the flux in g_2. For sure, this effect also plays a role for the a_1 and a_2 parameters, but the overall separation of normal and peculiar objects is much larger than +10 mmag (Figs. 72 and 73). In addition, also the g is, to a certain amount, affected by this decrease. Both parameters include the difference between g_2 and g which results in a certain diminishment.

For both parameters, a strong effect of the luminosity (or $\log g$) is apparent. One has to recall, that normally, giants can be well sorted out on the basis of a V or y versus $(g_1 - y)$ diagram for star clusters (Sects. 17.3 and 18.2).

The T_{eff} range of interest was divided in three parts for the further discussion. One has to recall that the 3σ limit for the classical Δa index for stars (V brighter than 19.0 mag) in the LMC (in one distinct field) is at ±12 mmag (Sect. 19.2). For Galactic open clusters (Sect. 17.2.1) it can even reach half this value.

- $T_{\mathrm{eff}} < 8000$ K: The underabundant models for the MS deviate from the solar abundant ones between 20 and 50 mmag for a_1 as well as

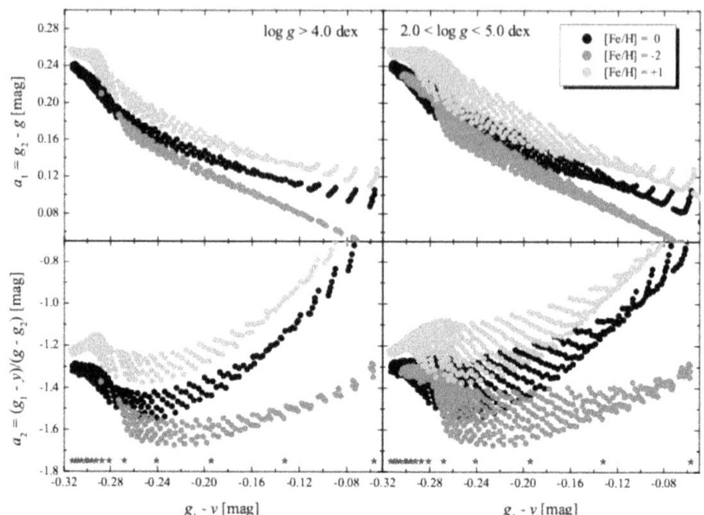

Figure 72: Two combinations of the Gunn g and g_1g_2y filters which would enhance the nowadays used Δa photometric system. The synthetic colours are for three different metallicities [Fe/H] =[−2,0,+1], and a microturbulent velocities ξ_t of $2\,\mathrm{km\,s^{-1}}$. The left panels show the MS whereas the right ones also include the effect of luminosity. The blue asterisks denote the corresponding $T_{\rm eff}$ values in steps of 1000 K starting with 6000 K at the red end of $(g_1 - y)$.

between 100 and 800 mmag for a_2. Even taking all $\log g$ values up to 2.0 dex, there are separated for the a_2 parameter.The overabundant models deviate by up to 10 mmag for a_1 and 100 mmag for a_2, respectively. Such a strong effect for cool type stars is not observed when looking at spectrophotometry (Sect. 9). Here, both parameters are much more superior than the a index alone.

- $8000 < T_{\rm eff} < 13000$ K: The underabundant models overlap with the solar abundant ones for both parameters, even taking only the MS values. However, in overall, the underabundant models tend to show lower values, or in other words, they represent the lower envelope. The overabundant models, are distinct by a constant value of about and 50 and 10 mmag, respectively. As a conclusion it can be said that both parameters do not supersede the capabilities of the classical Δa photometric system.

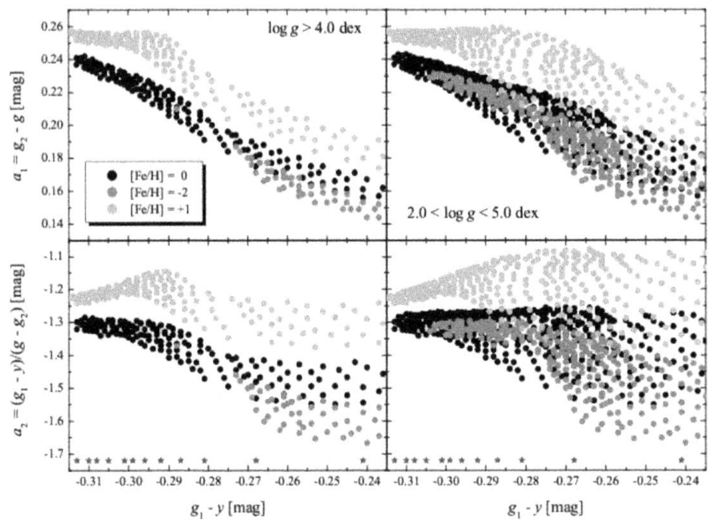

Figure 73: The same as Fig. 72, but for T_{eff} hotter than 9000 K.

- $T_{\text{eff}} > 13000$ K: Let's recall that this is the T_{eff} range for which the classical Δa photometric system is most sensitive (Sect. 11). This region comprises the (magnetic) CP2 and CP4 stars (Sect. 6). From Fig. 73, one can see that the underabundant models are not separated from the solar abundant ones. However, the overabundant ones are between 10 and 100 mmag, respectively. Furthermore, there is no overlap for higher luminosities models. This means that overabundant stars can be unambiguously detected with both parameters.

It can be concluded that additional observations, or available archival data, with the Gunn g would gain a significant improvement of the currently used Δa photometric system.

21 The ALHAMBRA and COSMOS surveys

In Sect. 8.4, spectral templates of different galaxy types without a redshift were presented. It was shown that the region from 4900 to 5700Å is well suited to study galaxies which would break new ground. This will be the next logical step in the evolution of the Δa photometric system after studying stars, star clusters, and the Magellanic Clouds. However, studying galaxies in a narrow-band photometric system, large telescopes are needed. Normally, it is not possible to use other filter sets than those provided by the corresponding observatory. Thus, using a g_2 filter at the VLT, for example, seems very unrealistic in the future.

Therefore, another approach was chosen in order to investigate galaxies (and also stars as it turned out) within this spectral region. The current available photometric surveys were searched for filters which correspond to the genuine g_2 one. Two surveys, namely ALHAMBRA and COSMOS, were found to include one filter fulfilling, within acceptable limits, the requirements. Both surveys are aimed at studying galaxies at high galactic latitudes ($|b| > 40^o$). Although the stellar densities at such high galactic latitudes is already quite low, also stars of all types are observed simultaneously with the galaxies. So it is possible to investigate galaxies and stars within the same homogeneous photometric system. This is an unique opportunity on the basis of high quality public available data sets.

The Advanced Large Homogeneous Area Medium-Band Redshift Astronomical (ALHAMBRA) survey (Moles et al., 2008) employs 20 contiguous, equal-width, medium-band filters covering from 3500Å to 9700Å, plus the standard JHK_s near-infrared (NIR) bands. A total area of four square degrees on the sky will be observed. In Fig. 74, the corresponding filters centred at 4570 and 5220Å are shown. The latter filter samples the 5200Å very well whereas the bluer one samples the continuum between Hγ and Hβ. The consecutive filter centred at 5510Å (not shown in this figure) samples the continuum redwards of the 5200Å without the influence of the Hα line. Molino et al. (2014) presented the public data release of this survey including about 100 000 galaxies, 20 000 stars in the galactic halo, and 1000 AGN candidates. The stars were identified according to their apparent geometry (PSF) on the image, F814W magnitude, as well as a combination of the optical (F489W − F814W) and NIR ($J - K_s$) colours (Troncoso Iribarren et al., 2016), respectively. However, up to now, no comprehensive analysis of the stellar sample was published. In a forthcoming project, the photometric data will be used to find faint and most distant CP stars on the basis of a system similar to the "classical" Δa one (Sect. 3). In addition, the galaxies observed by this survey will be also analysed in the same way as the stars.

The Cosmic Evolution Survey (COSMOS, Scoville et al., 2007) is designed to probe the formation and evolution of galaxies as a function of both cosmic time (redshift) and the local galaxy environment. The survey covers a two square degree equatorial field with imaging by most of the

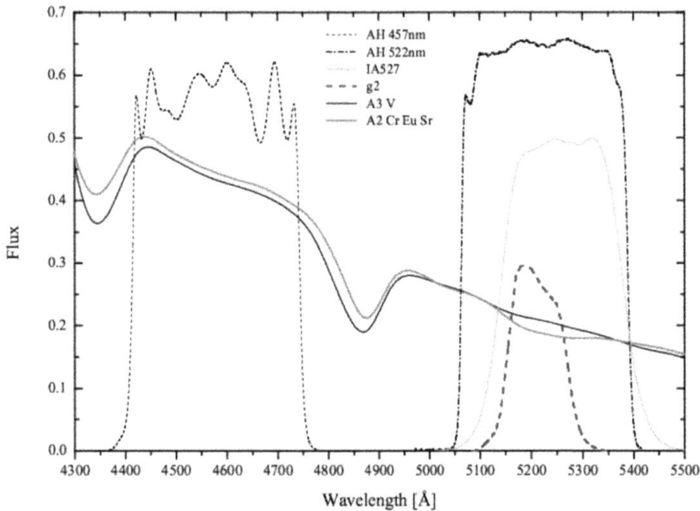

Figure 74: A comparison of the ALHAMBRA AH 457nm and AH 522nm (black dashed-dotted), Subaru Telescope Intermediate Band Filter IA527 (green), and g_2 (blue) filters together with spectrophotometric data of HD 102647 (A3 V; black) and HD 118022 (A2 Cr Eu Sr; red) taken from Section 9.

major space-based telescopes and a number of large ground based telescopes, with many ongoing surveys. The COSMOS survey is centred at (J2000): $\alpha = 10{:}00{:}28.6$ and $\delta = +02{:}12{:}21$. Objects as faint as 26th magnitude in the optical region are observed. Laigle et al. (2016) presented the "COSMOS2015 Catalog" which contains precise PSF-matched photometry, photometric redshifts and stellar masses for more than half a million of sources on the COSMOS field. It includes photometry in 36 different bands from the NUV (2314Å) to the FIR (79595Å). Among the whole variety of filters, it includes the "Subaru Telescope Intermediate Band Filter" IA527 which is centred at 5261Å with a FWHM of 242Å (Fig. 74). This filter is very useful in the context of studying the 5200Å region. From the huge amount of available filters, the most efficient ones will be chosen in order to establish an $a_{\rm Gal}$ index (Sect. 8.4), for example. This catalogue is already public available and will serve as a starting point for a pilot study of galaxies in the near future.

Although the observed fields of both surveys do not overlap, the final results will be compared and used to analyse the sample of galaxies.

22 Conclusions and Outlook

In this work it was shown how the Δa photometric system successfully produced countless scientific output and how it developed from a tool to detect CP stars to an universal tool to investigate even extragalactic systems. This enormous evolution was done within the last 40 years since its first appearance on the astronomical scene by Maitzen (1976a).

It has to be said that in the 1980ies and 1990ies, the system had its first period of prosperity because of the huge efforts by the the European working group on chemically peculiar stars of the upper main sequence (Mathys et al., 1989). Dedicated observing campaigns not only at the European Southern Observatory, but also at northern observatories such as at Hvar (Croatia) and Mt. Schöpfl (Austria) led to an active community and many publications. Other than, for example the Geneva 7-colour system, the characteristics of the Δa filter system slightly changed during time evolving to an almost perfect tool for detecting magnetic CP stars. At this time, the observations were devoted to Galactic field stars and open clusters, only. Joint efforts to shed more light on the CP phenomenon, to study magnetic fields, and the first important steps of asteroseismology made these research fields very vibrant.

However, at the end of the 1980ies, the highly developed photomultiplier technology, slowly but surely, was replaced by the CCD technology. It took a few years until CCDs reached the same accuracy as photomultipliers for the same exposure time and target brightness.

The second period of prosperity began with the first CCD Δa observations which were performed 1995 with the same telescope, the 61 cm Bochum telescope at La Silla, by the same person, Hans-Michael Maitzen, who initiated in 1971 the observations of CS Vir in g_1 and g_2. The results of the first CCD observations, data for 25 Galactic field stars with known spectroscopic peculiarity types, were published by Maitzen et al. (1997). After that, CCD observations of open clusters, the LMC and Globular clusters were successfully secured.

Other than in the photomultiplier era, no large international working group was conducting the observations and their analysis.

Nevertheless, this work has shown that the Δa photometric system and the study of its corresponding spectral region from 4900 to 5700Å is able to contribute significant new insights at many fields of astrophysics. The list of objects which are potential targets is long: cool-type Population I and Population II objects, supergiants, emission type objects, all type of galaxies, quasars, and so on.

These research fields can be investigated by 1) new observations; 2) archival data not used so far, for example cool-type objects in star cluster fields and 3) data from other surveys which employ similar filters. For sure, all three approaches will be extensively utilized in the future.

Several projects have already started or will soon be conducted. Here is a short list of these projects.

- Reduction of unpublished photoelectric and CCD observations of Galactic open as well as Globular clusters
- Execution of a complete survey of the Small and Large Magellanic Cloud
- Re-analysis of cool-type stars in star cluster fields
- Analysis of the public available data from the ALHAMBRA and COSMOS surveys in order to investigate galaxies and stars
- Observations of bona-fide Galactic field CP stars
- Observations of bright Abell galaxy clusters
- Observations of bright quasars
- Observations within the Gunn g filter and/or search for archival data to extend/upgrade the current Δa photometric system
- Spectroscopic observations of bona-fide photometric CP stars in the LMC and Globular clusters
- Comprehensive analysis of more sophisticated galaxy models taking into account redshifts

These projects will guarantee that the classical Δa photometric system will further evolve and become even more successful and important as it already currently is.

23 References

Abbott, T. M. C., & Kleinman, S. J. in Shafter A. W., ed, Interacting binary stars, ASP Conf. Ser., 56, San Francisco, p. 407

Abt, H. A. 1979, ApJ, 230, 485

Abt, H. A. 1985, ApJ, 294, L103

Abt, H. A., Brodzik, D., & Schaefer, B. 1979, PASP, 91, 176

Abt, H. A., Levato, H., & Grosso, M. 2002, ApJ, 573, 359

Adamczak, J., & Lambert, D. L. 2014, ApJ, 791, 58

Adelman, S. J. 1973a, ApJ, 182, 531

Adelman, S. J. 1973b, ApJ, 183, 95

Adelman, S. J. 1975, ApJ, 195, 397

Adelman, S. J. 1979, AJ, 84, 857

Adelman, S. J. 1980, A&A, 89, 149

Adelman, S. J. 1981, A&AS, 44, 265

Adelman, S. J. 1999, A&AS, 136, 379

Adelman, S. J., & Rayle, K. E. 2000, A&A, 355, 308

Adelman, S. J., & Wolken, P. R. 1976, ApJ, 207, 159

Adelman, S. J., Pyper, D. M., Shore, S. N., White, R. E., & Warren, W. H., Jr. 1989, A&AS, 81, 221

Adelman, S. J., Pyper, D. M., Lopez-Garcia, Z., & Caliskan, H. 1995, A&A, 296, 467

Adelman, S. J., Pi, C.-L. M., & Rayle, K. E. 1998, A&AS, 133, 197

Adelman, S. J., Caliskan, H., Cay, T., Kocer, D., & Tektanali, H. G. 1999, MNRAS, 305, 591

Adelman, S. J., Adelman, A. S., & Pintado, O. I. 2003, A&A, 397, 267

Ahumada, J. A., & Lapasset, E. 2007, A&A, 463, 789

Alcaino, G., Liller, W., Alvarado, F. et al. 1998, AJ, 115, 1492

Alecian, G. 2015, MNRAS, 454, 3143

Alecian, G., & Stift, M. J. 2010, A&A, 516, A53

Alves, D. R. 2004, New Astronomy Reviews, 48, 659

Anders, E., & Grevesse, N. 1989, Geochim. Cosmochim. Acta, 53, 197

Anderson, J., Sarajedini, A., Bedin, L. R. et al. 2008, AJ, 135, 2055

Andrievsky, S. M., Chernyshova, I. V., & Ivashchenko, O. V. 1995, A&A, 297, 356

Andrievsky, S. M., Chernyshova, I. V., & Kovtyukh, V. V. 1996, A&A, 310, 277

Antipova, L. I., Boyarchuk, A. A., Pakhomov, Yu. V., & Panchuk, V. E. 2004, Astronomy Reports, 48, 597

Asplund, M., Grevesse, N., Sauval, A. J., & Jacques, S. P. 2009, ARA&A, 47, 481

Babcock, H. W. 1947, ApJ, 105, 105

Babcock, H. W. 1958, ApJS, 3, 141

Balaguer-Núnez, L., Tian, K. P., & Zhao, J. L. 1998, A&AS, 133, 387

Balmforth, N. J., Cunha, M. S., Dolez, N., Gough, D. O., & Vauclair, S. 2001, MNRAS, 323, 362

Baumgardt, H., Dettbarn, C., & Wielen, R. 2000, A&AS, 146, 251

Bayer, C., Maitzen, H. M., Paunzen, E., Rode-Paunzen, M., & Sperl, M. 2000, A&AS, 147, 99

Behr, B. B. 2003a, ApJS, 149, 67

Behr, B. B. 2003b, ApJS, 149, 101

Behr, B. B., Cohen, J. G., McCarthy, J. K., & Djorgovski, S. G. 1999, ApJ, 517, L135

Bernhard, K., Hümmerich, S., Otero, S., & Paunzen, E. 2015, A&A, 581, A138

Bernhard, K., Hümmerich, S., & Paunzen, E. 2015, Astron. Nachr., 336, 981

Bernstein, R., Shectman, S. A., Gunnels, S. M., Mochnacki, S., & Athey, A. E. 2003, SPIE, 4841, 1694

Bertaud, C. 1970, A&AS, 1, 7

Bessell, M. S. 2005, ARA&A, 43, 293

Bidelman, W. P. 1985, AJ, 90, 341

Bidelman, W. P., & MacConnell, D. J. 1973, ApJ, 78, 687

Bigot, L., & Kurtz, D. W. 2011, A&A, 536, A73

Blauuw, A. 1965, in Galactic Structure, Blauuw A., Schmidt M., eds, University of Chicago Press, Chicago, p. 435

Bobylev, V. V., & Bajkova, A. T. 2014, MNRAS, 437, 1549

Böhm-Vitense, E., & Johnson, P. 1977, ApJS, 35, 461

Boesgaard, A. M. 1987, ApJ, 321, 967

Bonatto, C., Lima, E. F., & Bica, E. 2012, A&A, 540, A137

Borra, E. F., Joncas, G., & Wizinowich, P. 1982, A&A, 111, 117

Bowyer, S., Sasseen, T. P., Wu, X., & Lampton, M. 1995, ApJS, 96, 461

Breger, M., Beck, P., Lenz, P., Schmitzberger, L., Guggenberger, E., & Shobbrook, R. R. 2006, A&A, 455, 673

Brescia, M., Cavuoti, S., D'Abrusco, R., Longo, G., & Mercurio, A. 2013, ApJ, 772, 140

Brown, T. M., Sweigart, A. V., Lanz, T., Landsman, W. B., & Hubeny, I. 2001, ApJ, 562, 368

Buchholz, M., & Maitzen, H. M. 1979, A&A, 73, 222

Burnashev, V. I. 1985, Abastumanskaya Astrofiz. Obs., Byull., 59, 83

Bychkov, V. D., Bychkova, L. V., & Madej, J. 2003, A&A, 407, 631

Bychkov, V. D., Bychkova, L. V., & Madej, J. 2009, MNRAS, 394, 1338

Caliskan, H., & Adelman, S. J. 1995, in Adelman S.J., Wiese W.L., eds, Astrophysical Applications of Powerful New Databases, ASP Conf. Ser. 78, San Francisco, p. 443

Calzetti, D., Kinney, A. L., & Storchi-Bergmann, T. 1994, ApJ, 429, 582

Canto Martins, B. L., das Chagas, M. L., Alves, S. et al. 2011, A&A, 530, A73

Canuto, V. M., & Hartke, G. J. 1986, A&A, 168, 89

Canuto, V. M., & Mazzitelli, I. 1991, ApJ, 370, 295

Castelli, F. 1999, A&A, 346, 564

Castelli, F., & Kurucz, R. L. 1994, A&A, 281, 817

Castelli, F., Gratton, R., & Kurucz, R. L. 1997, A&A, 318, 841, erratum: 1997, A&A, 324, 432

Catalano, F. A., & Leone, F. 1994, A&AS, 108, 595

Catalano, F. A., & Renson, P. 1998, A&AS, 127, 421

Catalano, F. A., Leone, F., & Kroll, R. 1998, A&AS, 129, 463

Catalano, F. A., Leone, F., & Kroll, R. 1998, A&AS, 131, 63

Catelan, M. 2009, Ap&SS, 320, 261

Cenarro, A. J., Peletier, R. F., Sánchez-Blázquez, P. et al. 2007, MNRAS, 374, 664

Chen, B., Vergely, J. L., Valette, B., & Carraro, G. 1998, A&A, 336, 137

Chen, C. H., Mittal, T., Kuchner, M. et al. 2014, ApJS, 211, 25

Choudhury, S., Subramaniam, A., & Cole, A. A. 2016, MNRAS, 455, 1855

Chu, Y.-H., Gruendl, R. A., Stockdale, C. J. et al. 2004, AJ, 127, 2850

Chumak, Y. O., Platais, I., McLaughlin, D. E., Rastorguev, A. S., & Chumak, O. V. 2010, MNRAS, 402, 1841

Cidale, L. S., Arias, M. L., Torres, A. F. et al. 2007, A&A, 468, 263

Cioni, M.-R. L., Habing, H. J., & Israel, F. P. 2000, A&A, 358, L9

Clampitt, L., & Burstein, D. 1997, AJ, 114, 699

Clariá, J. J., Piatti, A. E., & Lapasset, E. 1998, A&AS, 128, 131

Claret, A. 1995, A&AS, 109, 441

Code, A. D., Bless, R. C., Davis, J., & Brown, R. H. 1976, ApJ, 203, 417

Cohen, R. E., Hempel, M., Mauro, F. 2015, AJ, 150, 176

Collins, G. W. II, Truax, R. J., & Cranmer, S. R. 1991, ApJS, 77, 541

Conti, P. S. 1976, in Slettebak A., ed., Be and Shell Stars, IAU Symposium no. 70, p. 447

Corbally, C. J. 1984, ApJS, 55, 65

Corliss, D. J., Morrison, N. D., & Adelman, S. J. 2015, AJ, 150, 190

Cowley, A. 1973, PASP, 85, 314

Cowley, A., Cowley, C., Jaschek, M., & Jaschek C. 1969, AJ, 74, 375

Cowley, C. R., Ryabchikova, T. A., Kupka, F. et al. 2000, MNRAS, 317, 299

Cowley, C. R., Hubrig, S., & González, J. F. 2009, MNRAS, 396, 85

Cowley, C. R., Hubrig, S., Palmeri, P. et al. 2010, MNRAS, 405, 1271

Cowley, C. R., Przybilla, N., & Hubrig, S. 2015, A&A, 578, A26

Cramer, N. 1984, A&A, 132, 283

Cramer, N. 1999, New Astronomy, 43, 343

Cramer, N., & Maeder, A. 1979, A&A, 78, 305

Cramer, N., & Maeder, A. 1980, A&A, 88, 135

Crawford, D. L. 1958, ApJ, 128, 185

Crawford, D. L. 1975, AJ, 80, 955

Crawford, D. L. 1978, AJ, 83, 48

Crawford, D. L. 1979, AJ, 84, 1858

Crawford, D. L., & Mander, J. 1966, AJ, 71, 114

Croom, S. M., Ratcliffe, A., Parker, Q. A. et al. 1999, MNRAS, 306, 592

Däppen, W. 1994-2000, private communication

Deutsch, A. J. 1970, ApJ, 159, 985

Dias, W. S., Alessi, B. S., Moitinho A., & Lépine, J. R. D. 2002, A&A, 389, 871

Dirsch, B., Richtler, T., Gieren, W. P., & Hilker, M. 2000, A&A, 360, 133

Dolphin, A. E. 2000, MNRAS, 313, 281

Domingo, A., & Figueras, F. 1999, A&A, 343, 446

Dubath, P., Rimoldini, L., Süveges, M. et al. 2011, MNRAS, 414, 2602

Dunstall, P. R., Fraser, M., Clark, J. S. et al. 2012, A&A, 542, A50

Dworetsky, M. M., Trueman, M. R. G., & Stickland, D. J. 1980, A&A, 85, 138

Eaton, J. A. 2008, AJ, 136, 1964

Elkin, V. G. 1995, Astron. Zh., 72, 879

Eggen, O. J. 1980, ApJ, 238, 627

Eigenbrod, A., Mermilliod, J.-C., Clariá, J. J., Andersen, J., & Mayor, M. 2004, A&A, 423, 189

Falcón-Barroso, J., Sánchez-Blázquez, P., Vazdekis, A. et al. 2011, A&A, 532, A95

Faraggiana, R., Bonifacio, P., Caffau, E., Gerbaldi, M., & Nonino, M. 2004, A&A, 425, 615

Feinstein, A. 1964, Observatory, 84, 11

Ferraro, F. R., Valenti, E., Straniero, O., & Origlia, L. 2006, ApJ, 642, 225

Finkenzeller, U., & Mundt, R. 1984, A&AS 55, 109

Francis, P. J., Hewett, P. C., Foltz, C. B., Chaffee, F. H., Weymann, R. J., & Morris, S. L. 1991, ApJ, 373, 465

Freyhammer, L. M., Elkin, V. G., Kurtz, D. W., Mathys, G., & Martinez, P. 2008, MNRAS, 389, 441

Frolov, V. N., Ananjevskaja, Yu. K., & Polyakov, E. V. 2012, Astronomy Letters, 38, 74

Garcia, B. 1993, ApJS, 87, 197

García-Gil, A., García L., Ramón, J., Allende Prieto, C., & Hubeny, I. 2005, ApJ, 623, 460

Gelbmann, M. 1998, PhD thesis, Univ. Vienna

Gerbaldi, M., Faraggiana, R., Burnage, R. et al. 1999, A&AS, 137, 273

Gerbaldi, M., Faraggiana, R., & Lai, O. 2003, A&A, 412, 447

Girardi, L., Groenewegen, M. A. T., Hatziminaoglou, E., & da Costa, L. 2005, A&A, 436, 895

Glagolevskij, Yu. V. 2005, Astronomy Reports, 49, 1001

Glagolevskij, Yu. V. 2013, Astrophysics, 56, 173

Glagolevskij, Yu. V., & Nazarenko, A. F. 2015, AstBu, 70, 89

Glebocki, R., & Stawikowski, A. 2000, Acta Astron., 50, 509

Glushneva, I. N., Kharitonov, A. V., Kniazeva, L. N., & Shenavrin, V. I. 1992, A&AS, 92, 1

Golay, M. 1974, Introduction to astronomical photometry, D. Reidel Publishing Co., Dordrecht

Golay, M. 1994, in The MK process at 50 years. A powerful tool for astrophysical insight, ed. Ch. Corbally, R. O. Gray, & R. F. Garrison, San Francisco, ASP Conf. Series, 60, 164

Gómez, A. E., Luri, X, Grenier, S. et al. 1998, A&A, 336, 953

Goudfrooij, P., Girardi, L., Rosenfield, P. et al. 2015, MNRAS, 450, 1693

Gough, D. O., & Tayler, R. J. 1966, MNRAS, 133, 85

Grankin, K. N. 2016, Astronomy Letters, 42, 314

Gratton, R. G., Carretta, E., & Bragaglia, A. 2012, A&A Rev., 20, 50

Gray, D. F. 2008, The Observation and Analysis of Stellar Photospheres, Cambridge University Press

Gray, D. F. 2010, AJ, 140, 1329

Gray, R. O. 1989, AJ, 98, 1049

Gray, R. O., & Corbally, C. J. 1993, AJ, 106, 632

Gray, R. O., & Corbally, C. J. 2002, AJ, 124, 989

Gray, R. O., & Corbally, C. J. 2009, Stellar Spectral Classification, Princeton University Press

Gray, R. O., & Garrison, R. F. 1989, ApJS, 69, 301

Gray, R. O., Corbally, C. J., Garrison, R. F., McFadden, M. T., & Robinson, P. E. 2003, AJ, 126, 2048

Griffin, R. E. M., & Griffin, R. F. 2011, Astron. Nachr., 332, 105

Grundahl, F., Vandenberg, D. A., & Andersen, M. I. 1998, ApJ, 500, L179

Grundahl, F., Catelan, M., Landsman, W. B., Stetson, P. B., & Andersen, M. I. 1999, ApJ, 524, 242

Guerrero, C. A., Orlov, V. G., Monroy-Rodríguez, M. A., & Borges Fernandes, M. 2015, AJ, 150, 16

Gustafsson, B., Bell, R. A., Eriksson, K., & Nordlund, A. 1975, A&A, 42, 407

Gustafsson, B., Edvardsson, B., Eriksson, K. et al. 2008, A&A, 486, 951

Guthnik, P., & Prager, R. 1914, Veröffentl. der königl. Sternwarte Berlin-Babelsberg, 1

Gutierrez-Moreno, A. 1975, PASP, 87, 805

Handler, G. 2013, in Oswalt T.D. et al. (eds), Planets, stars and stellar systems, Springer Science + Business Media Dordrecht, p. 207

Hanuschik, R. W. 1987, A&A, 173, 299

Harris, W. E. 1996, AJ, 112, 1487

Hauck, B. 1974, A&A, 32, 447

Hauck, B., & North, P. 1982, A&A, 114, 23

Hauschildt, P. H., Allard, F., & Baron, E. 1999, ApJ, 512, 3777

Havnes, O., & Conti, P. S. 1971, A&A, 14, 1

Havnes, O. 1974, A&A, 32, 161

Hearnshaw, J. B. 2014, The Analysis of Starlight Two Centuries of Astronomical Spectroscopy, Cambridge University Press

Heck, A., Mathys, G., & Manfroid, J. 1987, A&AS, 70, 33

Heiter, U. 2002, A&A, 381, 959

Heiter, U., Kupka, F., Paunzen, E., Weiss, W. W., & Gelbmann, M. 1998, A&A, 335, 1009

Heiter, U., Weiss, W. W., & Paunzen, E. 2002, A&A, 381, 971

Heiter, U., Kupka, F., van 't Veer-Menneret, C. et al. 2002, A&A, 392, 619

Hempelmann, A., & Schöneich, W. 1988, Astron. Nachr., 309, 97

Hensberge, H., & De Loore, C. 1974, A&A, 37, 367

Hensberge, H., Maitzen, H. M., Deridder, G. et al. 1981, A&AS, 46, 151

Hensberge, H., Weiss, W. W., Catalano, F. A., & Schneider, H. 1982, IBVS, 2170

Hensberge, H., Maitzen, H. M., Catalano, F. A. et al. 1986, A&A, 155, 314

Herbig, G. H. 1960, ApJS 4, 337

Hesser, J. E., Harris, W. E., Vandenberg, D. A. et al. 1987, PASP, 99, 739

Hilditch, R. W., Hill, G., & Barnes, J. V. 1983, MNRAS, 204, 241

Hill, G. M., & Landstreet, J. D. 1993, A&A, 276, 142

Hill, G. M., Gulliver, A. F., & Adelman, S. J. 2010, ApJ, 712, 250

Hoag, A. A., & Applequist, N. L. 1965, ApJS, 12, 215

Høg, E., Fabricius, C., Makarov, V. V. et al. 2000, A&A, 355, L27

Horch, E. P., Davidson, J. W., Jr., van Altena, W. F. et al. 2006, AJ, 131, 1000

Horne, J. H., & Baliunas, S. L. 1986, ApJ, 302, 757

Houk, N., & Swift, C. 1999, Michigan catalogue of two-dimensional spectral types for the HD Stars, Vol. 5, University of Michigan, Ann Arbor

Howell, S. B. 2006, Handbook of CCD Astronomy, Cambridge University Press

Hubrig, S., & Mathys, G. 1995, Comments Astrophys., 18, 167

Hubrig, S., North, P., & Mathys, G. 2000, ApJ, 539, 352

Hubrig, S., Le Mignant, D., North, P., & Krautter, J. 2001, A&A, 372, 152

Hui-Bon-Hoa, A., LeBlanc, F., & Hauschildt, P. H. 2000, ApJ, 535, L43

Hümmerich, S., Bernhard, K., Paunzen, E., Hambsch, F.-J., Bohlsen, T., & Powles, J. 2017, MNRAS, 466, 1399

Humphries, C. M., Jamar, C., Malaise, D., & Wroe, H. 1976, A&A, 49, 389

Iliev, I. Kh., Paunzen, E., Barzova, I. S. et al. 2002, A&A, 381, 914

Iliev, I. Kh., Budaj, J., Fenovcík, M., Stateva, I., & Richards, M. T. 2006, MNRAS, 370, 819

Jamar, C. 1977, A&A, 56, 413

Jamar, C. 1978, A&A, 70, 379

Jamar, C., Macau-Hercot, D., & Praderie, F. 1978, A&A, 63, 155

Järvinen, S. P., Hubrig, S., Schöller, M. et al. 2016, Astron. Nachr., 337, 329

Jaschek, C., & Jaschek, M. 1995, The Behavior of Chemical Elements in Stars, Cambridge University Press, Cambridge, p. 160

Jaschek, M., & Jaschek, C. 1960, PASP, 72, 500

Jaschek, M., & Jaschek, C. 1987, A&A, 171, 380

Jaschek, M., & Jaschek, C. 1990, The Classification of Stars, Cambridge University Press

Jaschek, M., Jaschek, C., & Arnal, M. 1969, PASP, 81, 650

Javakhishvili, G., Kukhianidze, V., Todua, M., & Inasaridze, R. 2006, A&A, 447, 915

Jenkner, H., & Maitzen, H. M. 1987, A&AS, 71, 255

Johnson, C. I., & Pilachowski, C. A. 2012, ApJ, 754, L38

Johnson, C. I., Kraft, R. P., Pilachowski, C. A., Sneden, C., Ivans, I. I., & Benman, G. 2005, PASP, 117, 1308

Johnson, H. L. 1958, Lowell Obs. Bull., No. 4, 37

Jordi, C., Carrasco, J. M., Fabricius, C., Figueras, F., & Voss, H. 2010, Highlights of Spanish Astrophysics V, Astrophysics and Space Science Proceedings, Springer-Verlag, Berlin, p. 147

Joshi, S., Mary, D. L., Chakradhari, N. K., Tiwari, S. K., & Billaud, C. 2009, A&A, 507, 1763

Kafka, S., & Honeycutt, R. K. 2003, AJ, 126, 276

Kamp, I., & Paunzen, E. 2002, MNRAS, 335, L45

Kato, K., & Sadakane, K. 1999, PASJ, 51, 23

Keller, S. C., Wood, P. R., & Bessell, M. S. 1999, A&AS, 134, 489

Khalack, V., LeBlanc, F., & Behr, B. B. 2010, MNRAS, 407, 1767

Khan, S. A., & Shulyak, D. V. 2006, A&A, 448, 1153

Khan, S. A., & Shulyak, D. V. 2006, A&A, 454, 933

Khan, S. A., & Shulyak, D. V. 2007, A&A, 469, 1083

Kharchenko, N. V., Piskunov, A. E., Schilbach, E., Röser, S., & Scholz, R.-D. 2013, A&A, 558, A53

Kharitonov, A. V., Tereshchenko, V. M., & Knyazeva, L. N. 1988, The spectrophotometric catalogue of stars, Nauka, Alma-Ata (USSR)

Khokhlova, V. L., Vasilchenko, D. V., Stepanov, V. V., & Romanyuk, I. I. 2000, Astronomy Letters, 26, 177

Kinney, A. L., Calzetti, D., Bohlin, R. C. et al. 1996, ApJ, 467, 38

Knigge, Ch. 2015, Ecology of Blue Straggler Stars, Astrophysics and Space Science Library, Springer-Verlag Berlin, 413, p. 295

Kochukhov, O., Khan, S. A., & Shulyak, D. V. 2003, A&A, 433, 671

Kodaira, K. 1969, ApJ, 157, L59

Kodaira, K. 1973, A&A, 26, 385

Kohoutek, L., & Wehmeyer, R. 1999, A&AS, 134, 255

Kozlowski, S., Onken, C. A., Kochanek, C. S. et al. 2013, ApJ, 775, 92

Kraus, M., Oksala, M. E., Cidale, L. S., Arias, M. L., Torres, A. F., & Borges Fernandes, M. 2015, ApJ, 800, L20

Krtička, J., Mikulášek, Z., Zverko J., & Žižňovský, J. 2007, A&A, 470, 1089

Krtička, J., Mikulášek, Z., Henry, G. W. et al. 2009, A&A, 499, 567

Krtička, J., Mikulášek, Z., Lüftinger, T. et al. 2012, A&A, 537, A14

Kubiak, M. 1990, Acta Astron., 40, 349

Kudryavtsev, D. O., Romanyuk, I. I., Elkin, V. G., & Paunzen, E. 2006, MNRAS, 372, 1804

Künzli, M., & North, P. 1998, A&A, 330, 651

Künzli, M., North, P., Kurucz, R. L., & Nicolet, B. 1997, A&AS, 122, 51

Kupka, F., Piskunov, N., Ryabchikova, T. A., Stempels, C., & Weiss, W. W. 1999, A&AS, 138, 119

Kurucz, R. L. 1970, Smithsonian Astrophys. Obs. Spec. Rep., 309

Kurucz, R. L. 1979, ApJS, 40, 1

Kurucz, R. L. 1992, Rev. Mexicana Astron. Astrofis., 23, 45

Kurucz, R. L. 1993, Kurucz CD-ROMs 1-13 (Cambridge: SAO)

Kurucz, R. L. 2005, Mem. Soc. Astron. It. Supp., 8, 14

Kuschnig, R., Ryabchikova, T. A., Piskunov, N. E., Weiss, W. W., & Gelbmann, M. J. 1999, A&A, 348, 924

Laigle, C., McCracken, H. J., Ilbert, O. 2016, ApJS, 224, 24

Lamers, H. J. G. L. M., Zickgraf, F.-J., de Winter, D., Houziaux, L., & Zorec, J. 1998, A&A, 340, 117

Lanz, T., & Hubeny, I. 2003, ApJS, 146, 417

Lanz, T., & Hubeny, I. 2007, ApJS, 169, 83

Lata, S., Pandey, A. K., Sagar, R., & Mohan, V. 2002, A&A, 388, 158

LeBlanc, F., Monin, D., Hui-Bon-Hoa, A., & Hauschildt, P. H. 2009, A&A, 495, 937

LeBlanc, F., Khalack, V., Yameogo, B., Thibeault, C., & Gallant, I. 2015, MNRAS, 453, 3766

Lehmann, H., Tkachenko, A., Fraga, L., Tsymbal, V., & Mkrtichian, D. E. 2007, A&A, 471, 941

Leckrone, D. S., Fowler, J. W., & Adelman, S. J. 1974, A&A, 32, 237

Lemke, M. 1997, A&AS, 122, 285

Liu, H., Li, S-L., Gu, M., & Guo, H. 2016, MNRAS, 462, L56

López Ariste, A., & Sainz Dalda, A. 2012, A&A, 540, A66

López-García, Z., & Adelman, S. J. 1999, A&AS, 137, 227

Lyubimkov, L. S. 1986, Izv. Krymsk. Astrof. Obs. 75, 155 (in Russian)

Maitzen, H. M. 1976a, A&A, 51, 223

Maitzen, H. M. 1976b, A&A, 60, L29

Maitzen, H. M. 1980a, A&A, 84, L9

Maitzen, H. M. 1980b, A&A, 89, 230

Maitzen, H. M. 1981, A&A, 95, 213

Maitzen, H. M. 1982, A&A, 115, 275

Maitzen, H. M. 1985, A&AS, 62, 129

Maitzen, H. M. 1993, A&AS, 102, 1

Maitzen, H. M. 1998, Hvar Observatory Bulletin, 22, 83

Maitzen, H. M., & Catalano, F. A. 1986, A&AS, 66, 37

Maitzen, H. M., & Floquet, M. 1981, A&A, 100, 3

Maitzen, H. M., & Hensberge, H. 1981, A&A, 96, 151

Maitzen, H. M., & Moffat, A. F. J. 1972, A&A, 16, 385

Maitzen, H. M., & Muthsam, H. 1980, A&A, 83, 334

Maitzen, H. M., & Pavlovski, K. 1987a, A&A, 178, 313

Maitzen, H. M., & Pavlovski, K. 1987b, A&AS, 71, 441

Maitzen, H. M., & Pavlovski, K. 1989a, A&A, 219, 253

Maitzen, H. M., & Pavlovski, K. 1989b, A&AS, 81, 335

Maitzen, H. M., & Rakos, K. D. 1970, A&A, 7, 10

Maitzen, H. M., & Segewiss, W. 1980, A&A, 83, 328

Maitzen, H. M., & Schneider, H. 1984, A&A, 138, 189

Maitzen, H. M., & Schneider, H. 1987, A&AS, 71, 431

Maitzen, H. M., & Vogt, N. 1983, A&A, 123, 48

Maitzen, H. M., & Wood, H. J. 1983, A&A, 126, 80

Maitzen, H. M., & Wood, H. J. 1977, A&A, 58, 389

Maitzen, H. M., Albrecht, R., & Heck, A. 1981, A&A, 96, 174

Maitzen, H. M., Seggewiss, H., & Tüg, H. 1978, A&A, 62, 191

Maitzen, H. M., Weiss, W. W., & Wood, H. J. 1980, A&A, 81, 323

Maitzen, H. M., Pavlovski, K., Catalano, F. A., Deridder, Gh., & Hensberge, H. 1986, A&AS, 65, 497

Maitzen, H. M., Weiss, W. W., & Schneider, H. 1988, A&AS, 75, 391

Maitzen, H. M., Paunzen, E., & Rode, M. 1997, A&A, 327, 636

Maitzen, H. M., Pressberger, R., & Paunzen, E. 1998, A&AS, 128, 537

Majaess, D. J., Turner, D. G., & Lane, D. J. 2007, PASP, 119, 1349

Malfait, K., Bogaert, E., & Waelkens, C. 1998, A&A 331, 211

Marconi, M., Molinaro, R., Ripepi, V., Musella, I. & Brocato, E. MNRAS, 428, 2185

Masana, E., Jordi, C., Maitzen, H. M., & Torra, J. 1998, A&AS, 128, 265

Massa, D. 1989, A&A, 224, 131

Massey, P. 2002, ApJS, 141, 81

Massi, F., Giannetti, A., Di Carlo, E. et al. 2015, A&A, 573, A95

Mateo, M. 1988, ApJ, 331, 261

Mathys, G. 1988, The Messenger, 53, 39

Mathys, G., Maitzen, H. M., North, P. et al. 1989, The Messenger, 55, 41

Mathys, G., Hubrig, S., Landstreet, J. D., Lanz, T., & Manfroid J. 1997, A&AS, 123, 353

Maury, A. 1897, Ann. Astron. Obs. Harvard Vol. 28, Part 1

Mermilliod, J.-C. 1982, A&A 109, 37

Mermilliod, J.-C., Mermilliod, M., & Hauck, B. 1997, A&AS, 124, 349

Michaud, G., Tarasick, D., Charland, Y., Pelletier, C. 1983, ApJ, 269, 239

Mikulášek, Z., Krtička, J., Henry, G. W. et al. 2010, A&A, 511, L7

Miroshnichenko, A. S., Gray, R. O., Vieira, S. L. A., Kuratov, K. S., & Bergner, Yu. K. 1999, A&A, 347, 137

Mobasher, B., Capak, P., Scoville, N. Z. et al. 2007, ApJS, 172, 117

Moles, M., Benítez, N., Aguerri, J. A. L. et al. 2008, AJ, 136, 1325

Molinaro, R., Ripepi, V., Marconi, M. et al. 2012, ApJ, 748, 69

Molino, A., Benítez, N., Moles, M. et al., 2014, MNRAS, 441, 2891

Möller, C. S., Fritze-v. Alvensleben, U., & Fricke, K. J. 1997, A&A, 317, 676

Molnar, M. R. 1975, AJ, 80, 173

Morgan, W. W., Keenan, P. C., & Kellman, E. 1943, An Atlas of Stellar Spectra, University of Chicago Press, Chicago

Mucciarelli, A., Origlia, L., Ferraro, F. R., Bellazzini, M., & Lanzoni, B. 2012, ApJ, 746, L19

Muciek, M., Gertner, J., North, P., & Rufener, F. G. 1984, IBVS, 2480

Musielok, B., Lange, D., Schoenich, W. et al. 1980, Astron. Nachr., 301, 71

Napiwotzki, R., Schoenberner, D., & Wenske, V. 1993, A&A, 268, 653

Neckel, Th., Klare, G., & Sarcander, M. 1980, A&AS, 42, 251

Nemravová, J., Harmanec, P., Kubát, J. et al. 2010, A&A, 516, A80

Nesvacil, N., Lüftinger, T., Shulyak, D. et al. 2012, A&A, 537, A151

Netopil, M. 2013, PhD thesis, Univ. Vienna

Netopil, M., Paunzen, E., Maitzen, H. M. et al. 2005, Astron. Nachr., 326, 734

Netopil, M., Paunzen, E., Maitzen, H. M. et al. 2007, A&A, 462, 591

Netopil, M., Paunzen, E., Maitzen, H. M., North, P., & Hubrig, S. 2008, A&A, 491, 545

Netopil, M., Paunzen, E., Heiter, U., & Soubiran, C. 2016, A&A, 585, A150

Niedzielski, A., & Muciek, M. 1988, Acta Astron., 38, 225

North, P. 1984, A&A, 141, 328

North, P., & Kroll, R. 1989, A&AS, 78, 325

North, P., Hauck, B., & Straižys, V. 1982, A&A, 108, 373

Oestreicher, M. O., Gochermann, J., & Schmidt-Kaler, T. 1995, A&AS, 112, 495

Olsen, E. H. 1979, A&AS, 37, 367

Olsen, E. H. 1980, A&AS, 39, 205

Osawa, K. 1965, Ann. Tokyo Astron. Obs. Ser. II, 9, 121

Pâris, I., Petitjean, P., Ross, N. P. et al. 2017, A&A, 597, A79

Paunzen, E. 2015, A&A, 580, A23

Paunzen, E., & Maitzen, H. M. 2001, A&A, 373, 153

Paunzen, E., & Maitzen, H. M. 2002, A&A, 385, 867

Paunzen, E., & Netopil, M. 2006, MNRAS, 371, 1641

Paunzen, E., & Vanmunster, T. 2016, Astron. Nachr., 337, 239

Paunzen, E., Weiss, W. W., Heiter, U., & North, P. 1997, A&AS, 123, 93

Paunzen, E., Duffee, B., Heiter, U., Kuschnig, R., & Weiss, W. W. 2001, A&A, 373, 625

Paunzen, E., Iliev, I. Kh., Kamp, I., & Barzova, I. 2002a, MNRAS, 336, 1030

Paunzen, E., Pintado, O. I., & Maitzen, H. M. 2002b, A&A, 395, 823

Paunzen, E., Pintado, O. I., & Maitzen, H. M. 2003, A&A, 412, 721

Paunzen, E., Kamp, I., Weiss, W. W., & Wiesemeyer, H. 2003, A&A, 404, 579

Paunzen, E., Netopil, M., Iliev, I. Kh. et al. 2005, A&A, 443, 157

Paunzen, E., Schnell, A., & Maitzen, H. M. 2005, A&A, 444, 941

Paunzen, E., Netopil, M., Iliev, I. Kh. et al. 2006, A&A, 454, 171

Paunzen, E., Heiter, U., Fraga, L., & Pintado, O. 2012, MNRAS, 419, 3604

Paunzen, E., Netopil, M., Maitzen, H. M. et al., 2014, A&A, 564, A42

Pavlovski, K., & Maitzen, H. M. 1989, A&AS, 77, 351

Pendl, E. S., & Seggewiss, W. 1975, in Weiss W.W., Jenkner, H., Wood, H., eds, Physics of Ap Stars, IAU Coll. 32, Vienna Observatory, p. 357

Percy, J. R. 2007, Understanding Variable Stars, Cambridge University Press, Cambridge

Perraut, K., Brandão, I., Mourard, D. et al. 2011, A&A, 526, A89

Pettersen, B. R. 1991, Mem. Soc. Astron. It., 62, 217

Pietrzyński, G., Graczyk, D., Gieren, W. et al. 2013, Nature, 495, 76

Piotto, G., Bedin, L. R., Anderson, J. et al. 2007, ApJ, 661, L53

Piotto, G., Milone, A. P., Anderson, J. et al. 2012, ApJ, 760, 39

Piskunov, N. E., & Kupka, F. 2001, ApJ, 547, 1040

Pöhnl, H., Maitzen, H. M., & Paunzen, E. 2003, A&A, 402, 247

Pojmanski, G. 2002, Acta Astron., 52, 397

Popescu, B., Hanson, M. M., & Elmegreen, B. G. 2012, ApJ, 751, 122

Porter, J. M., & Rivinius, Th. 2003, PASP, 115, 1153

Preston, G. W. 1974, ARA&A, 12, 257

Prevot, L., Andersen, J., Ardeberg, A. et al., 1985, A&AS, 62, 23

Prugniel, Ph., & Soubiran, C. 2001, A&A, 369, 1048

Prugniel, P., Vauglin, I., & Koleva, M. 2011, A&A, 531, A165

Quievy, D., Charbonneau, P., Michaud, G., & Richer, J. 2009, A&A, 500, 1163

Rasmussen, M. B., Bruntt, H., Frandsen, S., Paunzen, E., &Maitzen, H. M. 2002, A&A, 390, 109

Relyea, L. J., & Kurucz, R. L. 1978, ApJS 37, 45

Ren, S., & Fu, Y. 2013, AJ, 145, 81

Renson, P. 1978, A&A, 63, 125

Renson, P. 1991, Catalogue Général des Etoiles Ap et Am, Institut d'Astrophysique Université Liège, Liège

Renson, P., & Manfroid, J. 2009, A&A, 498, 961

Robin, A., & Crézé, M. 1986, A&A, 157, 71

Roby, S. W., & Lambert, D. L. 1990, ApJS, 73, 67

Röser, S., Demleitner, M., & Schilbach, E. 2010, AJ, 139, 2440

Ryabchikova, T. A. 2005, Astronomy Letters, 31, 388

Ryabchikova, T. A., Adelman, S. J., Weiss, W. W., & Kuschnig, R. 1997, A&A, 322, 234

Ryabchikova, T. A., Piskunov, N. E., Savanov, I., Kupka F., & Malanushenko, V. 1999a, A&A, 343, 229

Ryabchikova, T. A., Piskunov, N. E., Stempels, H. C., Kupka F., & Weiss, W. W. 1999b, Phys. Scripta, T83, 162

Ryabchikova, T., Leone, F., & Kochukhov, O. 2005, A&A, 438, 973

Salaris, M., Percival, S., Brocato, E., Raimondo, G., & Walker, A. R. 2003, ApJ, 588, 801

Santucci, R. M., Placco, V. M., Rossi, S. et al. 2015, ApJ, 801, 116

Sariya, D. P., Yadav, R. K. S., & Bellini, A. 2012, A&A, 543, A87

Savanov, I. S. 1996, Astr. Rep., 40, 196

Schaller, G., Schaerer, D., Meynet, G., & Maeder, A. 1992, A&AS, 96, 269

Schmidt-Kaler, Th. 1982, in Landolt-Börnstein New Series, Group VI, Vol. 2b, p. 453

Schneider, H. 1993, in Dworetsky M.M., Castelli F., Faraggiana R., eds, Peculiar versus Normal Phenomena in A-type and Related Stars, ASP Conf. Ser. 44, San Francisco, p. 629

Schnell, A., & Maitzen, H. M. 1994, Proc. 25th Meeting EWG CP Szombathely, p. 87

Schnell, A., & Maitzen, H. M. 1995, IBVS, 4175

Schönberner, D., Andrievsky, S. M., & Drilling, J. S. 2001, A&A, 366, 490

Schröder, C., & Schmitt, J. H. M. M. 2007, A&A, 475, 677

Schuster, W. J., & Nissen, P. E. 1989, A&A, 221, 65

Schuster, W. J., & Parrao, L. 2001, Rev. Mex. Astron. Astrofis., 37, 187

Scoville, N., Aussel, H., Brusa, M. et al. 2007, ApJS, 172, 1

Sharma, S., Pandey, A. K., Ogura, K. et al. 2006, AJ, 132, 1669

Shulyak, D., Tsymbal, V., Ryabchikova, T., Stütz, Ch., & Weiss, W. W. 2004, A&A, 428, 993

Shulyak, D., Krtička, J., Mikulášek, Z. et al. 2010, A&A, 524, A6

Simon, T. & Ayres, T. R. 2000, ApJ, 539, 325

Simunovic, M., & Puzia, T. H. 2016, MNRAS, 462, 3401

Sippel, A. C., Hurley, J. R., Madrid, J. P., & Harris, W. E. 2012, MNRAS, 427, 167

Skiff B. A., 2016, Catalogue of Stellar Spectral Classifications, VizieR Online Data Catalog (http://cdsarc.u-strasbg.fr/viz-bin/VizieR?-source=B/mk)

Smalley, B. 1993, A&A, 274, 391

Smalley, B., & Dworetsky, M. M. 1993, A&A, 271, 515

Smalley, B., & Dworetsky, M. M. 1995, A&A, 293, 446

Smalley, B., & Kupka, F. 1997, A&A, 328, 349

Smalley, B., Gardiner, R. B., Kupka, F., & Bessell, M. S. 2002, A&A, 395, 601

Sokolov, N. A. 2006, MNRAS, 373, 666

Soszyński, I., Poleski, R., Udalski, A. et al. 2008, Acta Astron., 58, 163

Sota, A., Maíz Apellániz, J., Morrell, N. I. et al. 2014, ApJS, 211, 10

Stankov, A., & Handler, G. 2005, ApJS, 158, 193

Stellingwerf, R. F. 1978, ApJ, 224, 953

Stępień, K., & Muthsam, H. 1987, A&A, 185, 225

Stetson, P. B. 1987, PASP, 99, 191

Stibbs, D. W. N. 1950, MNRAS, 110, 395

Stift, M. J., & Alecian, G. 2012, MNRAS, 425, 2715

Straižys, V., & Sviderskiene, Z. 1972, A&A, 17, 312

Straižys, V., & Žitkevičius, V. 1977, Soviet Astron., 21, 559

Straižys, V., Cernis, K., Vansevicius, V. et al. 1986, Vilnius Astronomijos Observatorijos Biuletenis, 75, 3

Strassmeier, K. G., Handler, G., Paunzen, E., & Rauth, M. 1994, A&A, 281, 855

Strömgren, B. 1964, Astrophys. Norv., 9, 333

Strömgren, B. 1966, ARA&A, 4, 433

Stryker, L. L. 1993, PASP, 105, 1081

Takeda, Y., Sato, B., & Murata, D. 2008, PASJ, 60, 781

Tanaka, K., Sadakane, K., Narusawa, S. et al. 2007, PASJ, 59, L35

Testa, V., Ferraro, F. R., Chieffi, A. et al. 1999, AJ, 118, 2839

Thuan, T. X., & Gunn, J. E. 1976, PASP, 88, 543

Tokovinin, A., Mason, B. D., Hartkopf, W. I., Mendez, R. A., & Horch, E. P. 2015, AJ, 150, 50

Torrejón, J. M., Schulz, N. S., Nowak, M. A., Testa, P., & Rodes, J. J. 2013, ApJ, 765, 13

Townsend, R. H. D., Owocki, S. P., & Howarth, I. D. 2004, MNRAS, 350, 189

Troncoso Iribarren, P., Infante, L., Padilla, N. et al. 2016, A&A, 588, A132

Tsymbal, V. V. 1996, in Adelman S.J. , Kupka F., Weiss W.W., eds, Model Atmospheres and Spectral Synthesis, ASP Conf. Ser., 108, San Francisco, p. 198

Valcarce, A. A. R., Catelan, M., & Sweigart, A. V. 2012, A&A, 547, A5

van den Ancker, M. E., de Winter, D., & Tjin A Djie, H. R. E. 1998, A&A 330, 145

Venn, K. A., & Lambert, D. L. 1990, ApJ, 363, 234

Vidal, C. R., Cooper, J., & Smith, E. W. 1973, ApJS, 25, 37

Viskum, M., Hernandez, M. M., Belmonte, J. A., & Frandsen, S. 1997, A&A, 328, 158

Vogt, N., & Faúndez, A. M. 1979, A&AS, 36, 477

Vogt, N., Kerschbaum, F., Maitzen, H. M., & Faúndez-Abans, M. 1998, A&AS, 130, 455

Vazdekis, A., Sánchez-Blázquez, P., Falcón-Barroso, J. et al. 2010, MNRAS, 404, 1639

Wade, G. A., Debernardi, Y., Mathys, G. et al. 2000, A&A, 361, 991

Wahlgren, G. M., & Hubrig, S. 2004, A&A, 418, 1073

Wahlgren, G. M., Bord, D. J., & Cowley, C. R. 2001, in Mathys G., Solanki S.K., Wickramasinghe D.T., eds, Magnetic Fields across the HR Diagram, ASP Conf. Ser., 248, San Francisco, p. 361

Watson, C. L. 2006, Society for Astronomical Sciences Annual Symposium, 25, 47 ("AAVSO International Variable Star Index"; VSX)

Wesselius, P. R., van Duinen, R. J., de Jonge, A. R. W. et al. 1982, A&AS, 49, 427

Wraight, K. T., Fossati, L., Netopil, M. et al. 2012, MNRAS, 420, 757

Young, A., & Martin, A. E. 1973, ApJ, 181, 805

Yuan, H. B., Liu, X. W., & Xiang, M. S. 2013, MNRAS, 430, 2188

Zacharias, N., Finch, C. T., Girard, T. M. et al. 2013, AJ, 145, 44

Zboril, M., North, P., Glagolevskij, Yu. V., & Betrix, F. 1997, A&A, 324, 949

Zejda, M., Paunzen, E., Baumann, B., Mikulášek, Z., & Liška, J. 2012, A&A, 548, A97

Zorec, J., & Royer, F. 2012, A&A, 537, A120

www.ingramcontent.com/pod-product-compliance
Ingram Content Group UK Ltd.
Pitfield, Milton Keynes, MK11 3LW, UK
UKHW022001190726
13853UKWH00004B/1667

9 783736 972834